Digital Integrated Circuit Design Using Verilog and SystemVerilog

Digital Integrated Circuit Design Using Verilog and SystemVerilog

Ronald Mehler

ELSEVIER

AMSTERDAM • BOSTON • HEIDELBERG • LONDON
NEW YORK • OXFORD • PARIS • SAN DIEGO
SAN FRANCISCO • SINGAPORE • SYDNEY • TOKYO
Newnes is an Imprint of Elsevier

Newnes

Newnes is an imprint of Elsevier
The Boulevard, Langford Lane, Kidlington, Oxford OX5 1GB, UK
225 Wyman Street, Waltham, MA 02451, USA

British Library Cataloguing in Publication Data
A catalogue record for this book is available from the British Library

Library of Congress Cataloging-in-Publication Data
A catalog record for this book is availabe from the Library of Congress

ISBN: 978-0-12-408059-1

For information on all Newnes publications
visit our website at http://store.elsevier.com/

Printed and bound in the United States of America

14 15 16 17 18 10 9 8 7 6 5 4 3 2 1

Table of contents

About the author

Ronald W. Mehler

Ronald Mehler is a professor of electrical and computer engineering at California State University, Northridge. Prior to joining the faculty of CSUN, he worked as an engineer for several companies, primarily designing digital-integrated circuits for aviation, telecommunications, and general-purpose computing applications. He has been designing integrated circuits with hardware description languages since 1988.

He holds a B.S. in electrical and computer engineering from the University of Wisconsin, an M.S. in electrical engineering from Texas A&M University, and a Ph.D. in electrical engineering from the University of Texas at Dallas.

Preface

This is a book about using Verilog and SystemVerilog to design digital-integrated circuits. It takes the readers from the most fundamental elements of digital design through the design of sophisticated components and interfaces. Included are guidelines for optimizing designs and creating robust, reliable systems.

Digital-integrated circuits are the electronic brains behind all modern electronics. Communications, computers, aviation, automobiles, consumer appliances, and much more: if it runs on electricity, it has digital-integrated circuits someplace in the background. All modern digital circuits are designed with a hardware description language, and Verilog/SystemVerilog is the engineer's choice for the majority of new designs.

Beyond simply a language reference manual, this book not only teaches the syntax of Verilog/SystemVerilog hardware description language, it teaches how to effectively use it to produce optimized circuits that will work the first time, every time. It contains little-understood information on asynchronous interfaces, a common source of failure in digital designs, and a guide to design partitioning to produce optimal designs.

While no prior exposure to any hardware description language is expected, readers should have some basic knowledge of Boolean algebra and electrical engineering fundamentals such as Ohm's law.

This book is based on courses taught by the author at California State University. The courses are themselves based on the author's 20 years of experience in private industry designing digital circuits prior to joining the CSU faculty.

Acknowledgments

The author would first and foremost like to thank Janice Mehler for finding and correcting innumerable errors and instances of generally poor writing. Without her editing efforts, this would be a much worse book.

Thanks are also due to Troy Wood at Synopsys, Inc., who provided wise suggestions and shepherded relevant passages through legal wickets and to Texas Instruments for allowing use of circuit images.

All trademarks and copyrights used herein are the property of their respective owners.

Chapter

Introduction

Modern digital circuits are designed at an abstract level using a hardware description language and logic synthesis. This book covers the use of the most popular such language, Verilog/SystemVerilog. The rest of this chapter presents some historical context for designing with Verilog and offers a brief overview in each chapter.

WHO SHOULD READ THIS BOOK

This book is intended for those who design, verify, or otherwise work with digital circuitry. It is expected that readers will have some familiarity with digital concepts such as Boolean logic and flipflops but no prior exposure to Verilog or any other hardware description language (HDL) is needed. A review of the fundamental digital concepts is included in Appendix B.

Verilog and SystemVerilog are equally useful for the design of field programmable gate arrays (FPGAs) and custom hardware devices. The techniques for designing both are covered in this book.

When used as a textbook, it is suitable for advanced undergraduate and beginning graduate courses in digital design.

HARDWARE DESCRIPTION LANGUAGES AND METHODOLOGY

Verilog is an HDL. SystemVerilog is a superset of Verilog that also includes numerous constructs that are useful for verifying designs but do not always have any meaning for circuit description. SystemVerilog is sometimes called a hardware design and verification language (HDVL) or just a hardware verification language (HVL) rather than an HDL.

HDLs provide a method of specifying the behavior of a design without specifying any implementation. They use programming language-like syntax to indicate the logical functions that are to be implemented. A page of Verilog hardware description can look a lot like a page of a C language computer program, as their syntaxes are similar, but their objectives are different. A computer program is a series of instructions that can be run on a suitable computer. An HDL specification of a design describes the functioning of a design that can be turned into a new machine. The former utilizes existing hardware to transform data. The latter is used to create new hardware. HDL design is not computer programming.

Using an HDL, a proposed new design can be encoded and the design verified before any hardware is constructed. Using an HDL allows designers to operate at a higher level of abstraction than previous design methodologies, providing a huge boost in efficiency and productivity.

Once an HDL design has been verified, the code can be turned from an abstract, technology-independent description into a technology-specific gate-level implementation. This transformation is accomplished through a highly automated process of logic synthesis. Several design automation companies make logic synthesizers that can be used to affect this step. Postsynthesis, there are several more steps that must be taken before the design will be ready for production.

Figure 1.1 shows a typical HDL design flow. A concept for a new design is, if economics warrant, turned into a design specification. Design engineers take this specification and turn it into an

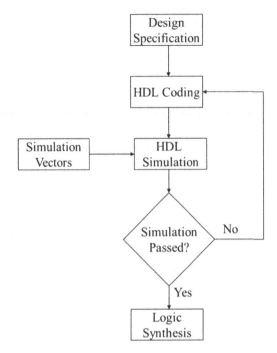

■ **FIGURE 1.1** HDL design flow

HDL description. In parallel, verification engineers write tests to determine if the HDL design implementation is correct, meets all the design specifications, and is sufficiently robust to operate under a variety of error conditions. Code written by the design team and the verification team is simulated. Once all agree that the HDL circuit description is complete and correct, the design is synthesized and turned into a gate-level netlist referencing a specific semiconductor technology.

WHAT THIS BOOK COVERS

This is a book about designing digital circuits with Verilog and SystemVerilog. It assumes no prior knowledge of Verilog or any HDL. It covers language syntax and best practices for producing reliable digital-integrated circuits. It includes hundreds of examples showing how the various constructs are used to effectively create hardware designs. It also includes numerous examples of test fixtures to verify the correct functioning of the provided design examples.

This book takes the reader from a design specification through a verified design ready for synthesis.

Covered are all parts of Verilog and SystemVerilog that are useful for circuit design as well as some components of the languages that are needed for verification but are not meaningful for circuit description.

SystemVerilog is an all-encompassing language that can be used for a variety of verification and even unrelated programming tasks that are not fundamental to hardware design. Constructs that are not directly related to hardware design are not covered.

HISTORICAL PERSPECTIVE

The earliest integrated circuits were designed at the subtransistor level. Individually crafting each transistor, a team of four engineers took 4 months to complete the first microprocessor. That processor, the four-bit Intel® 4004, used 2300 transistors. It went into production in 1971.

As this book is being written, some state of the art processors have upwards of three billion transistors. If engineers still work at the same rate of transistor design, a team of four would take nearly half a million years to turn out a new processor.

The density of transistors that could be formed on a single die increased exponentially over the past four decades as semiconductor manufacturing prowess improved. With the increase in transistor density, crafting each transistor individually became an untenable methodology. The necessity of developing ever-larger circuits led to the creation of more abstract models of combinational and sequential functions that could be reused in schematic representations of new circuits.

Designing at the gate level rather than at the transistor level and increasing the size of the design teams were the next steps in design methodology and management. These moves decreased the time to complete designs by orders of magnitude, but the inexorable increase in semiconductor density opened the doors to newer design methodologies. Unlike transistor density, design team size could not increase exponentially, year after year, for decades.

Verifying the behavior of these ever more complex circuit designs before committing to building the first prototype was another challenge. Simulations of the logical behavior of an abstract design became standard operating procedure, yet as design complexity continued to accelerate, determining if designs were logically correct became ever more difficult and time consuming.

It was to address verification challenges that what became HDLs were first developed. Building on earlier work with proprietary simulation languages, Philip Moorby and Prabhu Goel developed the first version of the Verilog language in the winter of 1983–1984 at their then-young startup company Gateway. At the same time, several companies were developing the earliest logic synthesis programs.

While it is implicit in the name (Verilog is formed from the words *veri*fy *log*ic) that the foci of efforts at Gateway were simulation and verification, the language's designers from the beginning were intent on using it for circuit specification and synthesis. Synopsys® was the first to license the new language for circuit synthesis from HDL code. At the time, it was a symbiotic relationship, as Synopsys did not then sell simulators and Gateway did not do logic synthesis.

Gateway was eventually bought by design automation company Cadence®, which does compete in the synthesis arena, and Synopsys has now long been in the simulation business as well as logic synthesis.

From its origins as a proprietary language, Verilog was released as an open standard in 1990 and in 1995 became an IEEE standard, IEEE 1364-1995. The standard has been updated and expanded several times, most recently by merging it with SystemVerilog, the object-oriented superset of Verilog. SystemVerilog is IEEE standard 1800.

When digital design with hardware description was new and engineers already had vast experience designing with schematic diagrams, there was some resistance to adopting the new HDL design paradigms. Early versions of the tools were far less capable than those of today. Fewer language constructs were supported and optimization was not as effective. At the beginning, there was some concern about the ability of automatic tools to generate correct gate-level circuits. Even once a high level of confidence that logic synthesis could produce logically correct circuits was obtained, engineers who had spent years and even decades optimizing gate-level designs remained convinced that they could obtain higher-quality results manually than any computer program could turn out.

Since there may have been some truth to that conviction when logic synthesis was first introduced, the HDL and synthesis design flow was first adopted by application-specific integrated circuit (ASIC) designers, for whom getting a completely new design into production quickly was more valued than squeezing the ultimate in clock speed out of a design. With a large foundation of highly optimized gate-level designs, processor developers were resistant to changing their fundamental design practices and procedures. While there was nothing inherently application specific about the HDL–synthesis design flow, it became known as ASIC methodology.

With improvements in design automation tools and under pressure to develop ever larger circuits with ever shorter design cycles, the ASIC methodology moved into less speed critical portions of processors and other catalog parts and then conquered the entire design process. Today using anything other than HDL and synthesis for digital design would be eccentric and anachronistic. It is used everywhere in the world where digital-integrated circuits are made and for products ranging from the simplest field programmable devices to the most advanced processors. HDL design is simply orders of magnitude more efficient than gate-level schematic design.

Competition-induced compressed design schedules coupled with the size of modern design have made older design methodologies impractical. HDL design and ASIC methodology is the way modern digital designs are made.

VERILOG AND SYSTEMVERILOG

This book covers Verilog and SystemVerilog for digital-integrated circuit design. While Verilog and SystemVerilog have been integrated into one IEEE standard, there are still many design automation tools in use that only support traditional Verilog as specified in the older IEEE Standard 1364. Accordingly, new SystemVerilog constructs are indicated as such, so that users of traditional Verilog can avoid use of unsupported features.

Since SystemVerilog is a superset of Verilog, anything that works in standard Verilog also works in SystemVerilog. When references are made to Verilog, users with access to SystemVerilog-compliant tools can assume that it applies equally to them.

The emphasis throughout this book is on the synthesizable subset of Verilog and SystemVerilog. Some nonsynthesizable constructs are included, as they are needed or at least useful for verifying synthesizable designs. This book does not provide comprehensive coverage of all of SystemVerilog, as a book providing that would be too large and unwieldy to be useful to design engineers.

BOOK ORGANIZATION

Chapter 1, this chapter, discusses the origins of HDL design and the imperative of using ASIC methodology for new designs. It stresses that, while hardware description code looks a lot like computer programming code, the two have entirely different purposes.

Chapter 2 presents the fundamentals of Verilog design. Using Boolean functions, it shows how logic gate type primitives, which are built into Verilog, can be organized into Verilog modules and complete designs. It covers single bit signals and organizing multiple bit signals into buses in Verilog. It introduces the most common Verilog variable types and hierarchical design, which is how large designs are constructed out of smaller modules. Also included are an introduction to verification and some examples of creating a Verilog test fixture to determine if a design module is working correctly.

Chapter 3 introduces behavioral HDL. It covers functional blocks, variable types, and behavioral operators. It also includes coverage of data structures and user-defined data types.

Chapter 4 continues coverage of behavioral coding by showing the different ways mnemonic aliases can be used, how user-defined elements can be shared across modules, and how to create scalable, reusable Verilog modules.

Chapter 5 covers loops and branches, which are among the more sophisticated behavioral coding techniques that give designing with HDL much of its power.

Chapter 6 explains and demonstrates the two types of subroutines available in Verilog, tasks and functions. It gives the rules for applying each and shows how SystemVerilog has both relaxed some of the rules and expanded subroutine capabilities. It also covers interfaces, a similar hierarchical construct used to simplify large designs.

Chapter 7 is on synchronization. Defective attempts at bridging clock domains are one of the primary reasons digital circuits fail. This chapter shows the techniques that work for creating robust asynchronous interfaces for different applications and conditions.

Chapter 8 deals with race conditions in both simulation and physical circuits. It explains the limitations of simulating parallel circuits on sequential computers and gives guidelines for designing reliable circuits.

Chapter 9 gives more guidance in creating high-quality designs. It covers design reuse, partitioning, and several types of optimization. It shows how architectural choices can have large impacts on circuit performance and how circuit size, speed, and power consumption can be traded off against each other to reach the most desirable circuit configuration for a particular application. It also shows some of the differences between FPGAs and ASICs and how to optimize the design for the targeted technology.

Chapter 10 is about test and testability. While these can be highly theoretical topics, the information provided here is intended to provide design engineers the knowledge needed to make designs testable without bogging down into the theoretical basis of testability. It explains why it is essential for designs to be made testable; despite the negative impact this will have on speed, area, and power consumption and how to do so. Covered are fault models, JTAG boundary scan, logic scan, built-in self-test, and parametric testing.

Chapter 11 is on hardware modeling. The models covered are Verilog representations of the physical cells a technology library will have. The output of the logic synthesis process is a netlist of these models, which can then be simulated prior to fabricating a circuit to ensure that timing constraints will not be violated and the physical device will work as desired. This chapter presents the knowledge needed to create and use cell models.

Chapter 12 includes several complete, synthesizable design examples as well as some verification modules for the presented designs.

Appendix A is a list of SystemVerilog reserved words.

Appendix B is a review of standard combinational and sequential functions.

Appendix C covers number systems commonly used in digital-integrated circuits.

Chapter **2**

Bottom-up design

This chapter introduces the most fundamental building blocks of a Verilog design. It shows how components can be used to build up a design hierarchy and includes several constructs needed to create and verify a design.

PRIMITIVE INSTANTIATION

While it largely defeats the purpose of working with a hardware design language (HDL), Verilog/SystemVerilog can be used to design any digital circuit using nothing but primitive Boolean operators. The standard logic types of AND, OR, NAND, NOR, XOR, XNOR, and NOT are all built into the language, as are buffers and Tri-state® drivers. There are four Tri-state driver types as shown in Table 2.1. The functioning of each primitive will be examined in this chapter.

These primitive operators can be instantiated to create any logic design, although their use is a tedious and error-prone exercise. Nevertheless, we will start with them, as it is a way to introduce Verilog syntax and structure while building on the fundamentals of digital logic that most readers are likely to be familiar with already.

Table 2.1 Tri-state driver functions

Verilog Name	Function When Control = 0	Function When Control = 1
bufif0	Output = input	Output is high impedance
bufif1	Output is high impedance	Output = input
notif0	Output = inverse of input	Output is high impedance
notif1	Output is high impedance	Output = inverse of input

Instantiation is a fundamental concept in HDL design, although it predates the invention of these languages. What it means is to place an abstract design object of some specified type in the design, where it can be connected to other instances and inputs or outputs. Instances are not physical components. When working with an HDL, any number of instances can be created and used simply by typing them into the design hierarchy.

Building designs through instantiating primitives is known as bottom-up design. Starting with the basic functions, the design is built up one instance at a time, ultimately creating the desired top-level design. While it is theoretically possible to create any conceivable design at all in this manner, doing so would be too time consuming to be practical for large designs.

When instantiating primitives, the size of each component is determined by the number of wires connected to the instance. There is no predefined relationship between an instance of a Verilog Boolean logic primitive and any physical gate of any technology. Gates of arbitrary size may be instantiated, regardless of the existence or nonexistence of any gate matching the instance parameters.

It is not only primitive operators that can be instantiated. Entire designs can be instantiated in other designs to create massive design hierarchies.

While primitives of any arbitrary size may be created, there are a few rules regarding their instantiation. As with any computer language, violation of syntactical rules will prevent the design from compiling and violations will have to be fixed with a text editor before proceeding.

The fundamental rules of Verilog primitives are that each instance must have exactly one output, unless the primitive is an inverter

(NOT) or buffer (BUF). In those cases, each instance must have exactly one input but may have an arbitrary number of outputs.

Primitive outputs must always come first in each instantiation, followed by as many inputs as the primitive has.

The built-in primitive types are Verilog keywords and may not be used for variable or design names. Like all Verilog keywords, they must be written entirely in lower case. User-defined primitives also may be created. Their names are not restricted to lower-case characters, although they must follow the other rules that apply to primitives. User-defined primitives will be covered in Chapter 11, Library Modeling.

Shown below is a Verilog instantiation of a three-input NAND gate.

```
nand   #3 G1 (OUT, A, B, C);
```

"nand" is a Verilog keyword specifying the type of instance. As is true of all Verilog keywords, it must be entirely written in lower-case characters. "#3" is an optional delay parameter associated with the instance. With this delay parameter, changes to the inputs will only be reflected at the output after three time units. The time units of this delay are defined in a timescale directive, which will be covered shortly. "G1" is the name of the instance. Instance names are also optional, although their use is recommended and is helpful in understanding and debugging designs. In parentheses are the wires connected to the instance. As always, the first one is the output. In this case, there are three inputs, A, B, C. The number of inputs is limited to 1024, effectively meaning that there is no limit. A semicolon ends the line. This punctuation is mandatory.

In this example, the wire and instance names are all written in upper case. While not a rule, many designers consider this to be good practice, as it helps distinguish variables from operators and other keywords.

An exhaustive simulation of the above gate instance is shown in Figure 2.1, along with a truth table.

All the fundamental Boolean operators may be instantiated in a similar fashion. As a review of Boolean fundamentals, a truth

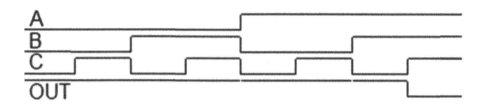

A	B	C	OUT
0	0	0	1
0	0	1	1
0	1	0	1
0	1	1	1
1	0	0	1
1	0	1	1
1	1	0	1
1	1	1	0

■ **FIGURE 2.1** Three input NAND gate simulation waveform and truth table

table for each of the built-in primitive type is shown below along
with a Verilog instance of each.

AND_OUT	A	B
0	0	0
0	0	1
0	1	0
1	1	1

and   A1(AND_OUT, A, B);

OR_OUT	A	B
0	0	0
1	0	1
1	1	0
1	1	1

or   O1(OR_OUT, A, B);

XOR_OUT	A	B
0	0	0
1	0	1
1	1	0
0	1	1

xor   X1(XOR_OUT, A, B);

NAND_OUT	A	B
1	0	0
1	0	1
1	1	0
0	1	1

nand   N1(NAND_OUT, A, B);

NOR_OUT	A	B
1	0	0
0	0	1
0	1	0
0	1	1

nor   N2(NOR_OUT, A, B);

XNOR_OUT	A	B
1	0	0
0	0	1
0	1	0
1	1	1

xnor   X2(XNOR_OUT, A, B);

NOT_OUT	A
1	0
0	1

not   N3(NOT_OUT, A);

BUF_OUT	A
0	0
1	1

buf   B1(BUF_OUT, A);

B_I_0	A	Control
0	0	0
Hi-Z	0	1
1	1	0
Hi-Z	1	1

bufif0 B2(B_I_0, A, Control);

B_I_1	A	Control
Hi-Z	0	0
0	0	1
Hi-Z	1	0
1	1	1

bufif1 B3(B_I_1, A, Control);

N_I_0	A	Control
1	0	0
Hi-Z	0	1
0	1	0
Hi-Z	1	1

notif0 B4(N_I_0, A, Control);

N_I_1	A	Control
Hi-Z	0	0
1	0	1
Hi-Z	1	0
0	1	1

notif1 B5(N_I_1, A, Control);

As already noted, each primitive must have exactly one output except the single-input primitives (not, buf). Those must have exactly one input. So the following instantiation

buf   BIGBUF(O1, O2, O3, A);

would have three parallel, identical outputs, all tracking the sole input, which is wire A. This feature is rarely useful, but it is legal.

DESIGNING WITH PRIMITIVES

The primitive operators can be instantiated any number of times and in any combination to create any digital design. To create a Verilog design, some overhead must be added. All Verilog circuit descriptions must be in a structure of "modules." A complete Verilog module inferring a 2:1 multiplexor along with a schematic representation of the same design is shown in Figures 2.2 and 2.3. A Verilog design is normally saved in a file having the same name as the module with a .v extension. SystemVerilog files use a .sv extension. While it would not be illegal to violate this norm, for example, saving a counter in a file called decoder.v, such obfuscation would not be accepted in any professional environment.

```
`timescale 1 ns / 1 ns
module MUX2_1(A, B, SEL, OUT);
  //Port declarations
  output OUT;
  input A, B, SEL;

  //Internal variable declarations
  wire SEL_N, A1, B1;

  //The netlist
  not (SEL_N, SEL);
  and (A1, A, SEL_N);
  and (B1, B, SEL);
  or  (OUT, A1, B1);

endmodule
```

■ **FIGURE 2.2** A complete Verilog module representing a 2:1 multiplexor

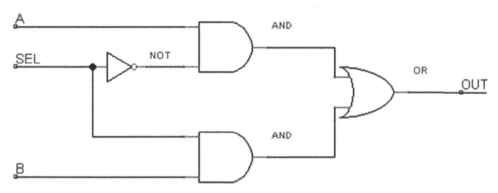

■ **FIGURE 2.3** Schematic representation of the 2:1 multiplexor described in Figure 2.2

The multiplexor module includes a timescale directive indicating that any delays in the module will be in units of 1 nanosecond and the smallest unit of time the simulator will track without rounding is also 1 nanosecond. Because the module does not contain any time delays, the timescale directive is without effect in this case.

Following the timescale directive, the keyword "module" appears. It is followed by the module name and, in parentheses, the module ports. Verilog ports are analogous to input, output, and bidirectional pins on a physical device.

Each port needs to have its direction specified. The options are input, output and inout, or bidirectional. This module is written in classic Verilog style. SystemVerilog allows a more compact version, declaring the direction of ports in the port list as shown in Figure 2.4.

```
`timescale 1 ns / 1 ns
module MUX2_1(input A, B, SEL, output wire OUT);

  wire SEL_N, A1, B1;

  not G1(SEL_N, SEL);
  and G2(A1, A, SEL_N);
  and G3(B1, B, SEL);
  or  G4(OUT, A1, B1);

endmodule
```

■ **FIGURE 2.4** SystemVerilog multiplexor module

In Figure 2.4, optional instance names are used for each primitive. The performance of the multiplexor models shown in Figures 2.2 and 2.4 would be identical.

After the port direction declarations, Figure 2.2 has further declarations of the internal wires that are used to connect the instances. By default, compilers are set to treat undeclared variables as single-bit wires, so the inclusion of these declarations is not always necessary. It is possible to change the default variable type to be something other than a single-bit wire or to disallow undeclared variables entirely. It is considered to be good practice to declare all variables before use.

Following the internal wire declarations are the instantiations of the primitives. Modules always end with the keyword "endmodule."

Even when working with primitives, design modification and maintenance are easy with Verilog. Consider now changing the multiplexor to add a third AND gate, one that will prevent switching glitches from appearing on the output, as shown in Figure 2.5.

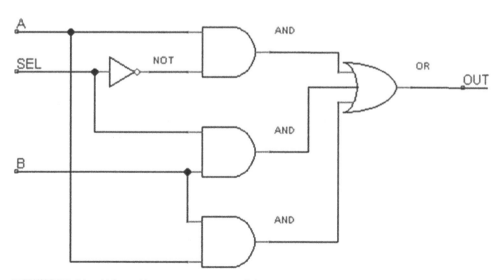

■ **FIGURE 2.5** 2:1 multiplexor with a gate to prevent output glitches

Prototyping this with hardware would require replacing a component and rewiring. With Verilog, however, making the change is nothing but a small editing job. The SystemVerilog code for the modified multiplexor is shown in Figure 2.6. Note that rather than replacing the OR gate, all that was necessary to do to it was to add another input to its port list.

```
`timescale 1 ns / 1 ns
module MUX2_1GK(input A, B, SEL, output wire OUT);

  wire SEL_N, A1, B1, AB;

  not G1(SEL_N, SEL);
  and G2(A1, A, SEL_N);
  and G3(B1, B, SEL);
  and G4(AB, A, B);
  or  G5(OUT, A1, B1, AB);

endmodule
```

■ **FIGURE 2.6** Glitch killing 2:1 multiplexor code

In these examples, several internal signals of type "wire" have been declared. The outputs also are declared to be type wire. Wires are one of the Verilog variable types. Verilog supports several hardware-oriented variable types that are not found in common programming languages. Other useful variable types will be introduced in later chapters.

When new variables are declared, each must be given a unique name. Verilog is case sensitive, whether applied to variables or keywords. Some examples of illegal, legal but ill-advised, and recommended wire declarations are shown in Figure 2.7.

```
Wire A, B; //Illegal, keywords must be all lower case
WIRE A, B; //Illegal, upper case again used in keyword
wire A, a; //Legal, but bad technique
wire a, b; //Legal, but may cause confusion
wire A, B; //The recommended way
```

■ **FIGURE 2.7** Legal and illegal wire declarations

While Verilog is case sensitive, not all design automation tools are. Thus, the declaration of two independent variables A and a as shown in Figure 2.7 would be legal, but could result in the two wires being shorted together later in the circuit development process. Many designers do use lower-case variables as shown in the second to last declaration, but a rigorous policy of capitalizing variables makes code easier to understand and maintain.

Verilog follows the common programming language convention of allowing comments anywhere in the code. Single line comments are preceded by double slashes. Multiline comments start with a /* and end with a */. An example of a multiline comment is shown in Figure 2.8.

```
module /*Comments may appear anywhere in a Verilog module*/ comment_module(A, B);
input A;
output B;
//Behavioral statements go here
endmodule
```

■ **FIGURE 2.8** Verilog multiline comment

IDENTIFIERS AND ESCAPED IDENTIFIERS

Variables, modules, and other constructs need names, more formally known as identifiers. In general, these names may not start with a numeral or use other nonalphabetic character at all, but that guideline can be finessed through the use of escaped identifiers.

The escape character is a backslash. Adding it allows otherwise illegal names to be used. Some sample-escaped identifiers are shown below. Their use is rare in modern designs. The examples below use reg variables as well as wires. Reg variables are covered in Chapter 3.

```
module \@4escapers ;
wire \5net ;
reg \mynet* ;
reg \negnet~ ;
wire \$dollar ;
```

In each of the above examples, there is white space between the escaped identifier and the line-ending semicolon. This is necessary. Without the white space, the semicolon would be incorporated into the escaped identifier and the compiler would not "see" the needed line termination. With a normal identifier, this white space is not needed.

Identifiers using numerals do not need to be escaped as long as they do not start with a numeral. The underscore character can be used in an identifier, although not as the first character. Other non-alphanumeric characters generally do need to be escaped when used in an identifier.

In the example below, ho_ho is legal without being escaped, but ho + ho needs to be escaped and to be followed by white space.

    reg ho_ho, \ho + ho ;

BUS DECLARATIONS

So far, all variables declared have been single-bit wires, but buses may also be declared. In the next example (Figure 2.9), the inputs are both eight-bit buses, while the output is again a single bit.

```
module WIDEGATE(input [7:0] A, B, output C);

  nor WG(C, A, B);

endmodule
```

■ **FIGURE 2.9** SystemVerilog multibit bus declaration

In Figure 2.9, a single instance of a 16-input NOR gate is inferred. It has exactly one output, the single-bit wire C. Note that the instance of the primitive follows the Verilog rule of having the output first, but the port order is changed in the module. This is not only legal but it is a conventional practice. Most designers prefer to put inputs first and outputs last in design modules. There is no rule on this, however, and it would be equally legal to mix up inputs and outputs, even deliberately obscuring the structure by alternating them or mixing them at random.

Another point to note in Figure 2.9 is that the declaration of A as an eight-bit bus carries through until the port type changes, so B is also an eight-bit bus. If the next input should be something different, another declaration can be made. In the following line, A will still be eight bits but B only one.

```
module WIDEGATE2(input [7:0] A, input B, output wire C);
```

The next example looks similar to Figure 2.9, but would be illegal. In Figure 2.10, the output is also set to be an eight-bit bus. Since there is still only one instance, that would be illegal. Every primitive instance other than buffers and inverters can only have a single output. Verilog does include a "generate" construct that could be used to create an array of single-output gates, but the code of Figure 2.10 would infer an impossible eight-output NOR gate.

```
//ILLEGAL!!! Will not compile.
module WIDEGATE(input [7:0] A, B, output wire [7:0] C);

   nor WG(C, A, B);

endmodule
```

■ **FIGURE 2.10** Illegal attempt to create multiple outputs with one instance

To make an eight-bit NOR function with primitives, eight instances are needed as shown in Figure 2.11. As already noted, this can also be accomplished with a generate loop. Generate loops and creating arrays of instances with them will be covered in Chapter 5.

```
module NOR8(input [7:0] A, B, output [7:0] C);

   nor G0(C[0], A[0], B[0]);
   nor G1(C[1], A[1], B[1]);
   nor G2(C[2], A[2], B[2]);
   nor G3(C[3], A[3], B[3]);
   nor G4(C[4], A[4], B[4]);
   nor G5(C[5], A[5], B[5]);
   nor G6(C[6], A[6], B[6]);
   nor G7(C[7], A[7], B[7]);
endmodule
```

■ **FIGURE 2.11** An eight-bit NOR function

Wires and other variable types may also be declared to be buses by putting their widths into brackets following the variable type, as shown below.

wire [15:0] DATA_BUS;

In this example, a 16-bit bus called DATA_BUS is declared and may subsequently be connected to suitable gates. Each bit of the bus can be individually referenced by placing its identifier in brackets, so to select the most significant bit (MSB) of DATA_BUS, all that would be needed to do is to reference DATA_BUS[15].

Verilog makes no assumptions as to big or little endian. Whatever appears to the left of the colon in a multibit declaration is the MSB. Whatever appears to the right is the least significant bit (LSB). In the following declaration, bit 0 is the MSB, bit 7 the LSB.

wire [0:7] BYTE_BUS;

That the MSB is the bit to the left of the colon in the declaration has implications for connecting buses or pieces of buses to other buses or pieces of buses. If one bus is declared to run from zero to 15 and another is from 15 to zero and they are then connected together, they will be connected MSB to MSB, LSB to LSB, not bit zero to bit zero, and 15 to 15.

If the buses are of unequal size, they will be matched from the LSBs unless explicitly instructed otherwise. For example, if BUS1 is 15 down to zero and is set to be equal to BUS2, which has zero as the MSB and seven as the LSB, they will be connected as shown in Figure 2.12. The upper bits of the larger bus will be left unconnected.

There is not even any requirement for bus declarations to go from zero to (size − 1), although that is the universally accepted standard. The following declaration creates a four-bit bus with 3 being the MSB and 6 the LSB.

wire [3:6] ODD_BUS;

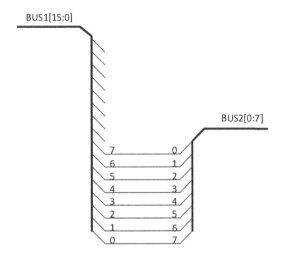

■ **FIGURE 2.12** Connecting a 15-bit little endian bus to an 8-bit big endian bus

DESIGN HIERARCHY AND TEST FIXTURES

Once a design has been completed, whether it is done with primitive instances or behavioral statements, it can be instantiated into a higher-level design or a test fixture. A design module containing one instance of the original multiplexor and one of the glitch-killing multiplexor is shown in Figure 2.13.

```
module muxes(input A, B, SEL, output wire [1:0] OUTS);
   MUX2_1    M0(A, B, SEL, OUTS[0]);
   MUX2_1GK  M1(A, B, SEL, OUTS[1]);
endmodule
```

■ **FIGURE 2.13** Upper-level module instantiating two design modules

In module MUXES, both designs share the inputs but have unique outputs. The two outputs are combined into a two-bit bus called OUTS, with bit zero of the bus the output of the original design and bit one the output of the modified glitch-killing module.

In the absence of gate delays, the two modules will work identically. However, if the gates are modeled with delays, they may simulate differently. Figure 2.14 shows two simulation runs, one without delays and the other having a 1-nanosecond delay

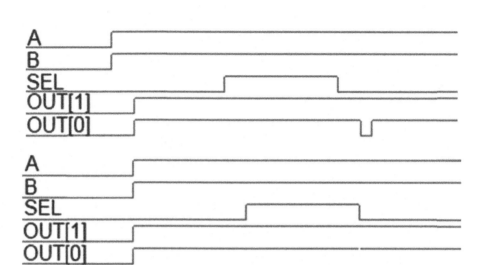

■ **FIGURE 2.14** Simulation of multiplexors with (top) and without (bottom) gate delays

associated with each gate. Without the glitch-killing gate, there will be a nanosecond of time when both A and B are equal to logic one, but both SEL and SEL_N are equal to zero, resulting in the output being zero. The glitch killer prevents this switching glitch and the output remains a constant logic one during the time that the change on SEL propagates through the circuit.

The engineer who can infallibly write correct HDL without any verification has not been and never will be born. In order to determine if a module is logically correct before any hardware is built, simulation is used. Before simulation can be run, a test fixture, also called a test bench, must be written. Verification is frequently more effort than creating the design in the first place, although the emphasis of this book is hardware design and the verification code will remain relatively simple.

A test fixture for the multiplexor of Figure 2.2 is shown in Figure 2.15. In this example, the inputs are stepped through all values from 000 through 111. This sort of exhaustive simulation is rarely practical in real designs, but can easily be done for this three-input device. In this example, a common discipline of naming the test

```
`timescale 1 ns / 1 ns
module tb_mux2_1;
  reg A, B, SEL;
  wire OUT;

  MUX2_1  UUT(A, B, SEL, OUT);

  initial begin
    #0 A = 1'b0; B = 1'b0; SEL = 1'b0;
    #1 A = 1'b0; B = 1'b0; SEL = 1'b1;
    #1 A = 1'b0; B = 1'b1; SEL = 1'b0;
    #1 A = 1'b0; B = 1'b1; SEL = 1'b1;
    #1 A = 1'b1; B = 1'b0; SEL = 1'b0;
    #1 A = 1'b1; B = 1'b0; SEL = 1'b1;
    #1 A = 1'b1; B = 1'b1; SEL = 1'b0;
    #1 A = 1'b1; B = 1'b1; SEL = 1'b1;
  end
endmodule
```

■ **FIGURE 2.15** Multiplexor test fixture, including instantiation of multiplexor design

bench tb_<module name> is used. Other conventions may be used, but having some consistent style is generally accepted practice.

The test fixture introduces a number of new constructs that are essential for any verification effort.

First, note that the test fixture has neither inputs nor outputs. It is entirely self-contained. This would be nonsensical for a hardware design module but is standard for the top-level test fixture.

Since the test fixture does have explicit delays, a timescale directive is used.

Following the module declaration, there are declarations of a new variable type, reg, and the now-familiar wire type. All the inputs are type reg, the output wire.

Reg is another hardware-specific type useful in HDL design and verification that is not used in common programming languages. The name would seem to imply that any variable declared to be a reg would be a register, that is, a flipflop, but such is not the case.

The name is anachronistic and dates to Verilog's origin as a verification language, predating hardware synthesis. The theoretical difference is that a variable of type reg retains its value until updated while a wire needs to be continuously driven to maintain a value. In practice, a reg variable may be the output of a combinational gate as well as a flipflop. We have already seen wires being used for the outputs of instantiated primitives, but the language makes no distinction on instantiated combinational or sequential components. If a flipflop were to be instantiated rather than a simple gate, its output would equally be connected to a wire.

A simple rule of thumb when writing test bench code is to make all inputs to the device under verification regs and all outputs or bidirectional signals wires. This rule of thumb works as long as the inputs are assigned their values inside of "initial" or "always" blocks, as is the case in this example and in the overwhelming majority of verification modules. Wire variables in the test fixture should be connected to output or bidirectional ports of the instantiated device under verification.

New SystemVerilog variable types add other options. When working with standard Verilog, the rule of thumb must always be used.

In Figure 2.15, the device to be verified is then instantiated. Note that although SystemVerilog allows the port type to be specified in the port list, when instantiated such identifiers must be stripped out.

Next comes an entirely new construct, an "initial" block. Initial blocks, along with always blocks, are the main vessels used in Verilog to contain behavioral statements. Both types of blocks may be used in test programs, but only always blocks are synthesizable and can be used in circuit descriptions.

The difference between an initial block and an always block is that initial blocks only run one time and always blocks may run an unlimited number of times. This accounts for the restriction against using initial blocks in circuit description code: hardware that takes an initial value and then can never change does not correlate to the way gates and flipflops actually work. Unlike syntax errors, including initial blocks in circuit description code will not cause code compilation to fail, but the initial blocks will be ignored in the synthesis process.

Although the test fixture of Figure 2.15 has only one initial block, any number of such blocks may be included. All start operating at simulation time zero, continuing to operate in parallel until each runs out of instructions.

In the initial block of Figure 2.15, the inputs are assigned values at time zero. Every 1 nanosecond thereafter, a new vector is applied to the inputs, eventually covering all eight binary combinations.

The syntax for each assignment has fields for the number of bits, the radix, and the value. Thus, the assignment A = 1'b0 means one bit of binary data with a value of zero should be placed on input A. The available radices are shown in Table 2.2.

Table 2.2 Verilog radices	
Radix	**Meaning**
b	Binary
o	Octal
d	Decimal
h	Hexadecimal

Unlike Verilog keywords, radix indicators may be upper or lower case. Thus, BYTEBUS = 8'HAA would assign the bit pattern 10101010 to the eight-bit variable BYTEBUS. Letters used in hexadecimal values also may be either upper or lower case; thus, BYTEBUS = 8'haa would be identical to the previous assignment, as would be BYTEBUS = 8'Haa and BYTEBUS = 8'hAa.

The initial block of Figure 2.15 is written for ease of understanding and maintainability, not minimal number of characters. Because all the inputs are regs, they do hold their values between assignments. Thus, the first two lines could equally be written

```
#0 A = 1'b0; B = 1'b0; SEL = 1'b0;
#1 SEL = 1'b1;
```

Since neither A nor B changes in the second vector, their previously assigned values are retained.

To save even more characters, the following syntax would also produce identical results:

```
#0 A = 0; B = 0; SEL = 0;
#1 SEL = 1;
```

This works because Verilog is not a strongly typed language. Numbers without size or radix are interpreted as decimal values. In a strongly typed language, one-bit variables could not be assigned decimal numbers, but Verilog allows it. Verilog is so loosely typed that numbers other than one and zero could also be used without ill effect. If the target of the assignment is smaller than the value assigned, Verilog will simply use the number of bits that do fit into the target and ignore the rest. Thus, the following assignments, while pointless and confusing, would also work exactly like the previous ones.

```
#0 A = 16'hfffE; B = 32'b10100110; SEL = 10;
#1 SEL =6'o35;
```

So far, all value assignments have, at a bit level, been nothing but ones and zeros. Verilog, however, is a quad-valued language. In addition to logic one and logic zero, variables can be in a high-impedance state or unknown. In Verilog, these conditions are indicated by Z and X, respectively. Like radix indicators, upper and lower case work identically.

Since X and Z could also be used as variable names, when used as values there must be some indicator that they are values and not variables. In the example below, Z would be a variable in the first line, but in the second, it is a value. In the first assignment, if

Z has not previously been declared to be a variable, a compilation error will result.

A = Z; //Z is a variable. It could have any value.
A = 1'bZ; //A is set to high-impedance.

Once low-level designs have been built and verified, they may be combined to form larger, hierarchical structures. In the next example, the assumption is made that a resettable D flipflop design has already been created and is going to be used to make other designs. The module statement for the flipflop is shown below, although code for its behavior is not included.

module DFFR(CLK, D, RST, Q);

Figure 2.16 then instantiates four of these flipflops to form a parallel four-bit register.

```
module REG4(input CLK, RST, input [3:0]
    Data, output wire [3:0] OUT);

    DFF D0(CLK, RST, Data[0], OUT[0]);
    DFF D1(CLK, RST, Data[1], OUT[1]);
    DFF D2(CLK, RST, Data[2], OUT[2]);
    DFF D3(CLK, RST, Data[3], OUT[3]);

endmodule
```

■ **FIGURE 2.16** Four flipflops instantiated to form a parallel register

There are four-bit input and output buses, with one bit of each connected to each D flipflop. Each instance is also given a unique name.

The same components can be slightly rearranged to form a four-bit shift register, as shown in Figure 2.16. In this example, simpler flipflops without a reset are used.

The sample circuit designs in Figures 2.16 and 2.17 both use D0, D1, D2, and D3 as instance names. Because instance names have to be unique, it may seem that instantiation of both SR4 and REG4 in a higher-level module could cause a conflict, but that is

```
module SR4(input CLK, SERIAL_IN, output wire SERIAL_OUT);

  wire STAGE1, STAGE2, STAGE3;

  DFF D0(CLK, SERIAL_IN, STAGE1);
  DFF D1(CLK, STAGE1, STAGE2);
  DFF D2(CLK, STAGE2, STAGE3);
  DFF D3(CLK, STAGE3, SERIAL_OUT);

endmodule
```

■ **FIGURE 2.17** Four flipflops organized into a shift register

not the case. The reason is that the instances with the same name are not in the same scope of the hierarchical design.

In Figure 2.18, designs SR4 and REG4 are themselves instantiated and each then has its own unique instance name. Thus, the path to D0 of the parallel register is X0.D0 and the path to D0 of the serial register is X1.D0. There is no ambiguity or conflict. In Verilog, the period is known as the "scoping operator." In a hierarchical design, an instance or signal at a lower level of hierarchy may be referenced by using this operator and instance names. In the design of Figure 2.18, the internal variable STAGE2 could be referenced by the hierarchical name X0.STAGE2 and the instance containing that wire is X0.D1. Instance names, not module names, are used when referencing hierarchical components. This is necessary because there may be many instances of a design and each would then have the

```
`timescale 1 ns / 1 ns

module TWOREGS(input CLK, RST, S_DATA, input [3:0]
  P_DATA, output S_OUT, output [3:0] P_OUT);

  SR4   X0(CLK, S_DATA, S_OUT); //Shift register instance
  REG4  X1(CLK, RST, P_DATA, P_OUT); //Parallel register instance;

endmodule
```

■ **FIGURE 2.18** Design instantiating hierarchical designs

same design name, but instance names are unique. Even if there is only one instance of a subdesign, attempting to reference it by design name would be an error.

Another point to note in Figures 2.16 and 2.18 is that the declarations of the module spill over into two lines. This is legal. The Verilog line is terminated by a semicolon. That it takes two lines to reach the punctuation mark is of no significance. Verilog, unlike some computer languages, does not have a continuation character.

PORT ASSOCIATION

Examples of design hierarchy so far have all used positional port association: when a module is instantiated, signals in the port list are associated with those in the instantiation in the order in which they appear. This is generally acceptable for small designs, but becomes awkward as the number of ports grows, especially when not all ports are connected.

Unconnected ports can be simply skipped in positional port association. An example of this is shown in Figure 2.19, where a flipflop with complementary outputs is instantiated twice but only one output of each instance is used. Numerous SystemVerilog behavioral constructs are used in the flipflop code, each of which will be examined in subsequent chapters.

Skipping a single unused output per instance in a design with only four ports is not onerous, but the situation changes when there are hundreds of ports with dozens of unused outputs. To help track what is connected where, named port association may be used.

In Figure 2.20, named port association is used when the flipflops are instantiated. In this example, the port order in each instance is in reverse order to that used in the flipflop model. With named port association, the order is not significant. Matching is done by names. No indicator at all is needed for unconnected ports in the instances, simplifying large designs and reducing the probability of connectivity errors.

```
module skip(CLK, RST, D, Q);
  input CLK, RST;
  input [1:0] D;
  output wire [1:0] Q;

  DFF R0(CLK, RST,, Q[0]); //QN is unconnected
  DFF R1(CLK, RST,, Q[1]);
endmodule

//Behavioral code modeling a D flipflop with
//complimentary outputs
module DFF(CLK, RST, D, QN, Q);
  input CLK, RST, D;
  output reg QN, Q;
  always_ff @(posedge CLK, negedge RST)
    if (!RST) begin
      QN <= 1'b1;
      Q <= 1'b0;
    end
    else begin
      QN <= ~D;
      Q <= D;
    end
endmodule
```

■ **FIGURE 2.19** Leaving an output port unconnected in a netlist

```
module named_assoc(CLOCK, RESET, DATA_INPUT, DATA_OUTPUT);
  input CLOCK, RESET;
  input [1:0] DATA_INPUT;
  output wire [1:0] DATA_OUTPUT;

  DFF R0(.Q(DATA_OUTPUT[0]), .D(DATA_INPUT[0]), .RST(RESET), .CLK(CLOCK));
  DFF R1(.Q(DATA_OUTPUT[1]), .D(DATA_INPUT[1]), .RST(RESET), .CLK(CLOCK));
endmodule
```

■ **FIGURE 2.20** Named port association

SystemVerilog adds wildcards to port association. With a wildcard, signals at the higher level will be automatically matched to signals of the same name in the instance. Signals that do not match can be associated via names. An example of this is shown in Figure 2.21, where all signal names match at both levels except for the clock. A limitation of this style is that port sizes must match. If there were two instances of each flipflop, as in the code of Figures 2.19 and 2.20, the D and Q ports would not match and would not be associated. Because the top-level design uses MASTER_CLK and the lower-level one uses CLK, the wildcard does not associate them and explicit named association must be used for that port.

```
module wildcard_assoc(MASTER_CLK, RST, D, Q);
  input MASTER_CLK, RST;
  input D;
  output Q;

  //Ports RST, D and Q will be covered by the wildcard
  DFF R0(.*, .CLK(MASTER_CLOCK));
endmodule
```

■ **FIGURE 2.21** Wildcard port association

When names match in the instance and the upper-level module, SystemVerilog offers another simplification, dot-name association. This style is similar to named port association, but the name in the upper-level module and the associated parenthesis can be left out. An example of this is shown in Figure 2.22. As with wildcards, the sizes of the signals must also match for association to be successful.

```
module dot_name_assoc(CLK, RST, D, Q);
  input CLK, RST;
  input D;
  output Q;

  //All instance ports will be matched to upper level
  //level ports with the same name.
  DFF R0(.Q, .D, .RST, .CLK);
endmodule
```

■ **FIGURE 2.22** Dot-name association

TIMESCALES

Timescales are directives used to instruct a Verilog simulator how to interpret time values embedded in code. In standard Verilog, timescales must appear outside of all modules. SystemVerilog adds some flexibility in timescale declarations.

In standard Verilog, timescale directives always start with a backtick (`` ` ``) followed by the word timescale, which oddly is not a Verilog keyword. Just to be perverse, it could be used as a variable or module name, although such perversity is seriously frowned upon in any professional environment.

Following `` `timescale `` are two numbers and the units associated with each. The first is the reference, which gives the multiplier factor for all timing data. The second is the precision. Any number less than the precision will be rounded to the nearest whole precision number. In the examples so far, these numbers and units have always been 1 ns, or one nanosecond, meaning that all timing data are to be given in units of 1 nanosecond and the simulator should round any numbers less than 1 nanosecond.

The only numbers legal for use in timescale directives are 1, 10, and 100. Any other number would cause a compilation error.

The legal timescale units are shown in Table 2.3.

Table 2.3 Timescale units

Symbol	Meaning
s	Seconds
ms	Microseconds (10^{-3} seconds)
us	Milliseconds (10^{-6} seconds)
ns	Nanoseconds (10^{-9} seconds)
ps	Picoseconds (10^{-12} seconds)
fs	Femtoseconds (10^{-15} seconds)

Another rule of timescales is that the precision cannot be less than the reference. Also note that, unlike Verilog statements, timescale directives do not end in a semicolon.

In the example of Figure 2.23, the reference is set to 100 picoseconds and the precision to 10 picoseconds. Because the delay of ANDGATE1 is specified to more precision than that specified in the timescale directive, it will be rounded. Changes on the inputs will be reflected 130 picoseconds later on the output.

```
`timescale 100 ps / 10 ps
module ANDMOD(input IN1, IN2, output OUT);
   and #1.26 ANDGATE1(OUT, IN1, IN2);
endmodule
```

■ **FIGURE 2.23** Timescale directive that will cause rounding

Figure 2.24 has some examples of legal and illegal timescale directives.

```
`timescale 1 ns / 10 ns //Illegal, reference is less than precision
`timescale 1 ns / 500 ps //Illegal, only 1, 10 and 100 are legal numbers
`timescale 1 ns / 100 ps //Legal
`timescale 100 ps /  10 ps //Legal
```

■ **FIGURE 2.24** Examples of legal and illegal timescale directives

Once a timescale has been set, it will be used until overridden by a new directive. This can cause some surprising results when multiple modules are compiled, some without timescales. In the three modules of Figure 2.25, module B will continue to use the 1 ns/1 ns timescale of module A, but only until the compilation order changes. If module C gets compiled before module B, the delays will suddenly become 1000 times greater than anticipated.

In SystemVerilog, the reference and precision can be split into two statements and embedded in a module as shown in Figure 2.26. Timeunit and timeprecision are SystemVerilog, but not standard Verilog, keywords. These instructions will override any previous timescale directives and stay in effect until overridden

```
`timescale 1 ns/1ns
module A(…);
  nand #3 (….)
endmodule
```

```
module B(…);
  nand #5 (….)
endmodule
```

```
`timescale 1 us/1us
module C(…);
  nand #0.03 (….)
endmodule
```

Compilation Orde

■ **FIGURE 2.25** Compilation order and timescales

```
module adder (input wire [63:0] A, B,
    output reg [63:0] SUM,
    output reg CARRY);
  timeunit 1ns;
  timeprecision 10ps;

  …
endmodule
```

■ **FIGURE 2.26** SystemVerilog timeunit and timeprecision statements

in turn. When using these new constructs, it is required that there be no space between the number and the unit. There is no such restriction with the older `timescale directive. This is shown in Figure 2.25, where some examples have a space between them and some do not. Such inconsistencies would be compilation errors with timeunit and timeprecision.

■ SUMMARY

The primitive operators presented in this chapter may be used to construct any circuit, but their use is tedious and error-prone. More advanced techniques that will vastly simplify the creation of circuits will be presented in subsequent chapters.

What will not change is the structure of modules. No matter what techniques are used, Verilog circuit descriptions need to be in modules, although as will be seen in later chapters, subroutines and other constructs may be used to facilitate the creation of flexible, reusable designs.

Hierarchical design and instantiation of lower-level modules will be used throughout the design and verification process.

Timescales should be used to ensure proper simulation but have no meaning in circuit descriptions.

Chapter

3

Behavioral coding part I: blocks, variables, and operators

All modern digital designs are done using behavioral hardware description language. This chapter introduces the essential elements to creating a design using code that describes what a design needs to do without specifying a detailed implementation. Operators, variables, and blocks are all covered. Operators are the symbols that indicate what manipulations are to be done. The different types of variables and when to use each are explained. Behavioral code is usually encapsulated in blocks, and the different options for blocks are also covered in this chapter.

TOP-DOWN DESIGN

The bottom-up methodology introduced in Chapter 2 can theoretically be used to create any digital design, but trying to make any significant design with only those tools would be too tedious and error-prone to be practical. Behavioral coding and top-down design are used for all modern designs.

When starting a new design from the top, the first steps are to define the overall functionality and then the inputs and outputs. These steps can be done without specifying any detailed implementation. For example, a processor design could be started by determining that there will be a 32-bit bidirectional combined address and data bus, a master clock, a reset, and a control bus consisting of a Read/Not Write signal, an interrupt input, and an address/not data signal. Then, without defining the processor further, the top-level Verilog template for the new machine could be coded as shown in Figure 3.1. As the design proceeds from top to bottom, the subblocks that will be instantiated in the top level and the behavior of these lower-level blocks will be added.

```
module TOPDOWN(input CLK, RST, INT, inout [31:0] BIBUS,
   output RD, AD);

   //instances of functional blocks go here

endmodule
```

■ **FIGURE 3.1** Top level of a machine that has not been fully defined

Once the behavior and interfaces for the blocks at the next level have been defined, coding these blocks can be assigned to different teams, with all the detailed work being done in parallel.

This top-down approach is the standard methodology for modern digital designs.

SYNTHESIZABLE AND NONSYNTHESIZABLE CODE

Verilog serves dual duty, as both a hardware description language and a verification language. Accordingly, there are some language constructs that are not applicable to circuit descriptions. While legal Verilog, these nonsynthesizable constructs must not be used in hardware descriptions. Circuit description is limited to the synthesizable subset of the language. In test fixtures, anything goes.

In addition to the inherently nonsynthesizable constructs, there are a few language elements that could theoretically be synthesizable but are not supported by synthesis tools. Real numbers fit into this category. Verilog supports real numbers, but no synthesizer currently does.

An example of inherently nonsynthesizable code would be an operation that infers a comparison to an unknown value, or a Verilog X. While the quad-valued language allows comparison to the value X, such a comparison is meaningless in hardware. X means indeterminate, that there is insufficient information to determine the value at that time. In hardware, however, a node that shows up as X in simulation would have a specific value when probed with an oscilloscope or voltmeter. This situation routinely occurs at the start of a simulation. Flipflops do not have a defined initial state. A physical flipflop, when first powered up, may output a logic zero or a logic one. It is indeterminate, but a real flipflop will be one or the other. In simulation, this ambiguity will result in register outputs starting out as X. Thus, comparisons to X can be done in test fixtures, but any such operations are non-synthesizable. Other nonsynthesizable constructs will be flagged as such as they are introduced.

REGISTER TRANSFER LEVEL (RTL)

Behavioral code for hardware design is normally written in the style known as Register Transfer Level, or RTL. With RTL style coding, the logic of a circuit is expressed with the various Verilog operators and the output of logic expressions are assigned, or transferred, to registers. Using this style, the registers, or flipflops, in a design are inferred with clock edge statements.

Circuit design examples throughout this book are written in RTL. Simply creating an "always" block that includes an edge specifier causes registers to become the object of behavioral code. Examples of clock edge statements are shown below.

```
always_ff @(posedge CLOCK, negedge RESET) //SystemVerilog style
always @(posedge CLOCK, negedge RESET) //Verilog 2001 style
always @(posedge CLOCK or negedge RESET) //Verilog 95 style
```

Each of the above edge specifiers infers rising edge-triggered registers with an asynchronous reset. If no reset is required, the following line would suffice:

```
always_ff @(posedge CLOCK)
```

Again, it is assumed that the registers will be active on the rising edge of the clock. Falling edge-sensitive flipflops can be inferred as well. The following line would be used in negative edge-triggered RTL:

```
always_ff @(negedge CLOCK)
```

In all of these examples, it is assumed that there are signals called CLOCK and RESET. These are not reserved words. Any legal Verilog identifiers can be used as the object of edge statements.

Always blocks are one of the two main types of functional blocks used in Verilog. Functional blocks are discussed in this chapter following the material on continuous assignments.

CONTINUOUS ASSIGNMENTS

The simplest types of behavioral constructs are continuous assignments. These statements start with the keyword "assign," followed by an expression that can be translated into a Boolean function. Continuous assignments are written outside of functional block structures, which are discussed later in this chapter.

Just as it is possible to design any digital device using nothing but two-input NAND gates, even the most complex machines may be designed using nothing but continuous assignments. This is never done, however, as continuous assignments provide only a marginal improvement over instantiating primitive operators. They are widely regarded as analogous to "goto" statements in programming: legal constructs that are to be avoided in well-structured code.

In addition to making the overall structure of a design obscure, continuous assignments are limited in that they cannot take advantage of Verilog high-level language constructs such as loops and multiway branching. A design consisting of nothing but continuous assignments leaves much of the power of HDL untapped.

A further argument against use of continuous assignments is simulation speed. Simulators must reevaluate continuous assignments at every instant in time, consuming more processing power than alternative coding techniques. Since any given node is, statistically speaking, likely to change state in less than 5% of the clock cycles, use of continuous assignments has a substantial performance penalty. Better simulation performance is achieved through the use of "always" and "initial" functional blocks, discussed in the following sections.

Nevertheless, continuous assignments are a part of the language and they are used in the examples that follow.

An adder can be constructed with a continuous assignment as shown below. This example uses an addition operator rather than being constructed out of nothing but Boolean operators. The carry out of the eight-bit operation is integrated into the SUM variable, creating a nine-bit result.

```
input [7:0] DATA1, DATA2;
output wire [8:0] SUM;
assign SUM = DATA1 + DATA2;
```

An even worse design of an adder can be made by limiting the continuous assignment statements to Verilog Boolean operators. In the following example, a one-bit full adder is created with the sum and carry variables separated so that each can be the target of a single continuous assignment.

```
input A, B, CARRY_IN;
output wire SUM, CARRY_OUT;
assign SUM  = A^ B ^ CARRY_IN;
assign CARRY_OUT = A & B | A & CARRY_IN | B & CARRY_IN ;
```

Designing at such a low level of abstraction would be no more effective than using nothing but primitive instantiations. In neither case could a design of any complexity be expected to be completed in any reasonable time frame. In the above example, the symbols for the Boolean operations XOR (^), AND (&), and OR (|) are used. They will be covered in more detail later in this chapter.

The one exception to the general policy of avoiding continuous assignments is when connecting signals to a bidirectional interface. Because of the Verilog resolution function, or the lack of a resolution function for some data types, use of continuous assignments on Tri-state bidirectional ports is a standard technique. This will be covered in more detail later in this chapter.

IMPLICIT CONTINUOUS ASSIGNMENTS

Instead of using the keyword "assign," a continuous assignment can be inferred when a variable is created. Thus, inferring the eight-bit adder of the example above could equally be accomplished by the following code:

```
input [7:0] DATA1, DATA2;
output wire [8:0] SUM = DATA1 + DATA2;
```

Embedding the function in the declaration has no impact on simulation performance or any circuit synthesized from the code. If this style is more or less obscure and convoluted than using an extra line of code to define the function is a matter of opinion on which the author takes no position. Avoidance of continuous assignments, whether implicit or explicit, is still the accepted methodology.

FUNCTIONAL BLOCKS: ALWAYS AND INITIAL

Other than continuous assignments and subroutines (Chapter 6), all Verilog behavioral code takes place inside of "initial" or "always" functional blocks. SystemVerilog adds extensions to always statements by allowing the always statement to clarify what type of hardware the block is to infer: combinational, latched, or registered. This is done by replacing the keyword "always" with always_comb, always_latch, or always_ff.

In simulation, all types of blocks start operating at time zero. Procedural blocks, whether initial or always, all run concurrently. The difference is that initial blocks only run once whereas always blocks restart every time they are triggered by a change on an input to the block. What this means in effect is that circuit descriptions may only use always blocks, as this corresponds to the way hardware actually works. Initial blocks are useful for simulation but are not synthesizable, as circuit components that stop responding to inputs after the initial values are exhausted does not correlate to anything that can be built.

A module can have any number of functional blocks. In a design, some blocks will infer combinational logic, some sequential.

Creating blocks is simply a matter of using the keywords initial or always, followed by the relevant behavioral statements. A block containing more than one Verilog statement must also use the keywords "begin" and "end," as shown in Figure 3.2 or "fork" and "join." Fork and join are nonsynthesizable and must not be used in circuit descriptions, although they are useful in verification programs.

```
module BLOCKS;

  reg CLK, A, B;

  initial begin
    //non-synthesizable statements
  end

  always_ff @(posedge CLK) begin
    //synthesizable SystemVerilog sequential statements
  end

  always_comb begin
    //synthesizable SystemVerilog combinational statements
  end

  always_latch begin
    //synthesizable SystemVerilog behavioral statements with
    //latched outputs
  end

  always @(*) begin
    //Verilog combinational statements
  end

  always @(A or B) begin
    //Verilog 95 style combinational statements
  end

endmodule
```

■ **FIGURE 3.2** Verilog and SystemVerilog functional block statements

The difference between fork...join and begin...end is that all statements in a fork...join block are run in parallel, whereas begin...end statements are evaluated sequentially. An application of the fork...join construct is shown in Figure 5.15, where parallel evaluation is needed for clock generators, but it is not hardware design.

Fork...join has three variants. Standard Verilog only supports fork...join. SystemVerilog has expanded this parallel construct to allow the use of join_any or join_none in place of join. The difference is the scheduling of statements following the parallel fork block. With a simple join, all tasks initiated in the fork block must end before anything following the parallel statements will run. With join_any, statements following the parallel block will start running as soon as any task in the parallel block finishes. With join_none, statements following the parallel block will be executed immediately following issuing the parallel statements and without waiting for any of them to complete. The differences will only become apparent when the parallel statements in a fork...join construct are routines that will take some time to complete. If they are nothing but simple register assignments that run in zero simulation time, all will appear identical.

Because no fork...join construct is synthesizable, these differences in order of execution on the host computer have no meaning for circuit design. They are only for use in verification programs.

Figure 3.2 contains examples of blocks written to take advantage of SystemVerilog advances as well as traditional Verilog coding styles.

The first block, an initial block, is nonsynthesizable. These typically are only used in test fixtures. If one is used in a circuit description, it will be ignored by the synthesizer. This can lead to simulation-synthesis mismatches, a potentially catastrophic situation. When there are such mismatches, the design that is verified in simulation will be different from that produced by synthesis.

While advanced designers sometimes deliberately do this to optimize hardware, it is a risky procedure.

The second block in Figure 3.2 would infer flipflops for all block outputs. This coding style is not available in classic Verilog. This type of sequential block requires a sensitivity list, a list of edge constructs that says at which events the block should be restarted and its outputs reevaluated.

The next block is a SystemVerilog combinational block. It does not need any explicit sensitivity list. It will be reevaluated whenever any inputs to the block change.

The always_latch construct signals that the block should create logic in which the outputs are latched. However, having such a block successfully infer the desired logic depends on the code in the block. Using the SystemVerilog keyword always_latch allows design automation tools to check that the code in the block does indeed infer latches and issue warnings if it does not, but does not guarantee that such logic will be created. The coding style necessary to generate latches will be covered in the section on multiway branching in Chapter 5.

Another way of inferring combinational logic is always (*). This means that the block should be reevaluated whenever any input to the block changes. This construct was added to Verilog in the 2001 update to the language standard.

The last case is the classic Verilog method of creating always blocks. It says to update evaluation of the block only when signals in the sensitivity list change. Because of frequent mistakes by designers in leaving signals out of the sensitivity list, this style has largely gone out of fashion, but it is still supported by all design automation tools.

A typical use of initial blocks is to simulate primary inputs to a device under verification. For example, a clock generator can easily be created with an initial block to provide a time base for the design. The periodic signal created in such a block can then be connected to the clock input of the design. An example of a complete clock generator block with a period of two is shown below.

```
initial begin
  CLK = 1'b1;
  forever #1 CLK = ~CLK;
end
```

NAMED BLOCKS

A functional block may be given a name, as shown in Figure 3.3. Named blocks can have local variables, and they can also use variables declared in the module but outside of the block. If a block has a local variable that has the same name as a module variable, references to that variable name in the block will be to the local variable and not to the module variable. There is no conflict, as by giving a block a name, a new level of hierarchy is created. In simulation, a named block may be disabled by name and any simulation events scheduled but not yet executed by that block will be removed from the execution queue. This has no hardware implications. Disabling a block may be done in verification but has no meaning in a circuit.

```
module namedblock;

  reg [3:0] COUNT; //module variable COUNT

  always_comb begin: BLOCK1
    reg [7:0] COUNT; //BLOCK1 variable COUNT
  end : BLOCK1

  always_comb begin
    //This block will use the module variable COUNT
  end

endmodule
```

■ **FIGURE 3.3** Named block with a local variable

Naming blocks create new levels of hierarchy. In Figure 3.3, the local variable COUNT inside of BLOCK1 could be referenced in the module by giving a hierarchical path to it, BLOCK1.COUNT.

In Figure 3.3, the named block ends with the block name repeated. This is done just to clarify the code and adds no functionality. Ending with ": BLOCKNAME" is optional. Neither the colon nor the block name is required.

Named blocks must use keyword pairs begin…end or fork…join even if they only contain a single behavioral assignment.

SENSITIVITY LISTS

In classic Verilog, evaluation of an always block can only be triggered by changes to elements in the block's sensitivity list. A sensitivity list can be either combinational or sequential. While simulators tend to allow mixed combinational and sequential sensitivity lists, synthesizers do not. For circuit description, a sensitivity list must be either combinational or sequential, never both.

A sensitivity list is made sequential by the inclusion of edge constructs. There are only two such constructs, posedge and negedge. Thus,

```
/*rising edge triggered flipflop(s) with asynchronous reset*/
always @(posedge CLK, negedge RST) begin
```

would infer flipflops clocked on the rising edge of signal CLK with an asynchronous, active low reset called RST.

A block without edge constructs in the sensitivity list will represent either purely combinational circuitry or the block outputs could be latched. A simple example could infer a single OR gate as follows:

```
always @(A or B) C = A | B; // A two-input OR function
```

This tells the simulator that it should update the function on any change of either A or B. An easy mistake to make with this coding style is to leave a signal out of the list, which can make the simulation of the design appear to have a memory effect. If B were left out of the sensitivity list in the above example, changes on B without a change on A would not cause output C to change. The memory effect is shown in Figure 3.4. Output Y, the one that is the result of a function with an incomplete sensitivity list, does not change when B changes. It appears to store previous values.

```
module BADLIST(input A, B, output reg Y, Z);
  always @(A) Y = A ^ B; //XOR, defective sensitivity list
  always @(A, B) Z = A ^ B; //XOR, complete sensitivity list
endmodule
```

Time 0: A = 0, B = 0, Y = 0, Z = 0
Time 1: A = 0, B = 1, Y = 0, Z = 1
Time 2: A = 1, B = 0, Y = 1, Z = 1
Time 3: A = 1, B = 1, Y = 1, Z = 0

■ **FIGURE 3.4** Defective sensitivity list and memory effect in simulation

Synthesizers ignore combinational sensitivity lists. The list is only checked to determine if the logic generated should be combinational or sequential. Thus, while the two outputs of example 3.4 simulate differently, they would synthesize to exactly the same hardware. This is an example of a simulation-synthesis mismatch that can occur with an incomplete sensitivity list.

Since the 2001 language update, elements in sensitivity lists may be separated by either commas, as shown in Figure 3.4, or the word "or," as shown in the last block of Figure 3.2. In earlier versions of Verilog, only "or" was accepted. These are the only options. Use of any other punctuation mark, conjunction, or logic type in a sensitivity list would be a syntax error.

Another innovation of the 2001 update was the * sensitivity list. Using that option, a combinational block will be reevaluated whenever any input to the block changes. Its use prevents the error of Figure 3.4. An application of the * in a sensitivity list is shown in the second to last block of Figure 3.2.

A sensitivity list mixing edge-sensitive signals with other variables would fail at compile time. The following is an example of this defective, illegal coding style. While simulators do allow it to pass, a circuit description including it would fail in synthesis even after simulating as intended.

always @(posedge CLK, RST) //ILLEGAL, will fail in synthesis

SPLITTING ASSIGNMENTS

It may seem like a logical simplification to split assignments to a variable between two functional blocks, as shown in Figure 3.5. This is always a mistake. It would infer a short circuit.

```
module shortcicuit(input CLK, RST, D, output reg Q);
/*Design error: this would appear to work in simulation
but would infer two flipflops shorted together in hardware.*/
   always @(negedge RST) Q <= 1'b0;
   always @(posedge CLK) Q <= D;
endmodule
```

■ **FIGURE 3.5** Two flipflops with outputs shorted together

This is another situation where simulation and synthesis would diverge. In simulation, the last assignment to Q at any given time would be the one displayed. There would be no indication of a malfunctioning device. However, synthesis would infer two flipflops, one negative-edge triggered connected to RST and one positive-edge connected to CLK. The two flipflop outputs would then be shorted together. Making assignments to non-Tri-state variables in multiple functional blocks is never the answer to any design problem, even if Verilog code doing precisely that appears to simulate correctly.

VARIABLES

Verilog supports common programming language data types such as integer, real, and string variables, and also includes several more that are useful in describing hardware behavior.

In standard Verilog, variable types are divided into two categories: nets and registers. The names are somewhat misleading, in that the variables in the register category, including type "reg," are not necessarily outputs of flipflops. Conversely, nets may be connected to flipflop outputs. This dates back to Verilog's beginnings as a verification language, before it was used for hardware description. The distinction was used to differentiate between variables that would hold their values between being updated and those that needed to be continuously driven. In properly structured hardware description code, reg variables may equally be the outputs of combinational gates, flipflops, or latches.

A further distinction is that variables in the net group have resolution functions and registers do not. The effect of this rule is that nets can have multiple drivers, which is the situation with Tri-state buses. It is not possible to make a Tri-state bus with nothing but register variables.

The most used variable types are wires, which are a type of net, and regs, which are a type of register. It is possible to go an entire career using nothing but these two variable types, but others can be useful.

A few designers prefer to declare Tri-state nets as type "tri," which work exactly as do variables of type wire. The point of using "tri" is simply to have the code self-document.

Nets

The Verilog net types are shown in Table 3.1. While legal Verilog, net types other than wire are rarely seen. The other net types are generally not supported for synthesis. Wand stands for Wired AND. Open collector logic families work in this manner: Outputs connected together act as if they were connected via an AND gate. Similarly, long-obsolete emitter-coupled logic devices were a Wired OR technology. Gate outputs could be connected together and would act as if they were all inputs to an OR gate.

Table 3.1 Verilog net types	
Net Type	**Function**
wire	Normal interconnects between instances
tri	Exactly the same as wire
wand	Wired AND. Models open drain/open collector devices
triand	Exactly the same as wand
wor	Wired OR. Used in now-obsolete emitter-coupled logic
trior	Exactly the same as wor
trireg	Models nets with capacitive storage. Holds last-driven value
tri0	Nets that pull up when not driven
tri1	Nets that pull down when not driven

Verilog has a rigid rule on where net variables may be assigned values. They may never be assigned values inside of functional blocks (always or initial blocks). Their use is limited to being the target of continuous assignments and interconnects between instances of primitives or instantiated modules.

Multibit net variables may be declared by giving a signal range along with the declaration. A delay may also be associated with a net, but delays are ignored by synthesizers. Depending on a delay for proper operation of a circuit could thus lead to simulation-synthesis mismatch.

A sample declaration of a five-bit bus with a six time-unit delay is shown below.

```
wire [4:0] #6 SLOWBUS;
```

By default, an undeclared variable will be a single-bit wire. This can be changed through use of the `default_nettype directive. The most common use of this directive, which must be used outside of any module, is to force all variables to be explicitly declared. This is done by making the default type "none," as shown below:

```
`default_nettype none
```

The reason for doing that is to prevent an overlooked variable declaration from becoming a design error. If two multibit variables are to be connected via a bus but the bus is undeclared, the result will be to have just one bit connected and all the other bits unconnected. If all variables are required to be declared, this error will be flagged at compile time.

Any of the other net types could be used as the default, but in any case whatever is selected must remain a single-bit type. There is no way to make a multibit bus the default type. The "none" option is only available in SystemVerilog. It does not work with standard Verilog.

Net aliases

Nets can be given multiple names through the use of aliases. The procedure is simply to make multiple declarations of nets and then assign them to each other.

The following lines of code allow the reset signal to be referenced by several different names. Only one net will be created. A value assigned to any of the reset name variations will cause all to take on that value.

```
wire RESET;
wire RST;
wire Reset;
wire Reset_n;

alias RST = RESET;
alias Reset = RST;
alias Reset_n = Reset;
alias NRESET = RESET; //NRESET was not declared, defaults to single bit wire
```

Bus variables too can be aliased. The following lines would create two names for one eight-bit bus. The order does not matter. Aliasing B_BUS to A_BUS would produce exactly the same results as the code shown.

```
wire [7:0] A_BUS;
wire [7:0] B_BUS;

alias A_BUS = B_BUS;
```

Net types can only be aliased to other nets of the same type and width. Only net types can have aliases. Any attempt to alias register variables would be an error. The following examples would all be syntax errors. Register variables are covered in the next section of this chapter.

```
wire [1:0] BUS;
wire A;
wand B;
wor C;
reg D;
alias A = BUS; //Error: Widths must match
alias B = C; //Error: Net types must match
alias D = A; //Error: Regs cannot be aliased
alias A = D; //Error: Wires cannot alias to regs
```

One practical application for aliases is to extract pieces of a bus. In the following example, a nine-bit bus has its least significant bit aliased to a parity variable, which can then be conveniently referenced for further processing.

```
input [8:0] DATA_BUS;
alias PARITY = DATA_BUS[0];
```

Net signal strength

Verilog nets can have a signal strength parameter. The possible values are shown in Table 3.2. They may be associated with gate instances. Drive strength specifications are ignored for synthesis.

Some examples of strength specifications are shown below. In simulation, conflicts will be resolved by having the stronger strength override the weaker. In a circuit, the result will be a short circuit and probably physical damage to the device.

```
// 6 strength for logic 1 output, 3 strength for logic 0 output
and (strong1, weak0)g1(Z, A, B);
// 7 strength for logic 1 output, 5 strength for logic 0 output
buf (pull0, supply1) g2(Z, A);
```

Table 3.2 Signal strengths

Strength Name	Value
supply1	7
strong1	6
pull1	5
large1	4
weak1	3
medium1	2
small1	1
highz1	0
highz0	0
small0	1
medium0	2
weak0	3
large0	4
pull0	5
strong0	6
supply0	7

When logic values are the same but strengths differ, the output will take the strength of the stronger driver. If strengths are the same but logic values differ, the output will be indeterminate. This holds true for values other than highz. It is illegal to connect together two or more nets all having the strengths of (highz1, highz0).

Supply0 and supply1 can also be used to declare nets that are hard wired to power and ground. In synthesis, such nets will be replaced with continuous assignments to logic one and logic zero. In the example below, net A will be continuously driven to logic one. The gate output Z will be the inverse of B.

```
supply1 A;
xor g1(Z, A, B);
```

Registers

Within procedural blocks, register variables are used in standard Verilog. In addition to variables of type reg, Verilog supports several real and integer data types, as shown in Table 3.3.

Table 3.3 Standard Verilog register data types	
Register Type	**Definition**
reg	Integer variable, may be signed or unsigned
integer	Signed integer, defined to be at least 32 bits. Actual implementation may vary.
real	64-bit floating point
realtime	64-bit floating point, identical to real
time	64-bit unsigned integer

Floating point numbers are unsupported for synthesis.

Integers are synthesizable, but their vague definition makes their use problematic. The implementation is required to be at least 32 bits, but it may be more. Hardware designers tend to use reg vectors rather than integers to avoid the implementations being dependent on tool versions.

Time variables are used for tracking simulation time. While they could be used in a circuit description, common practice is again to declare a 64-bit reg vector when such a variable is needed in a circuit.

The assignment rules for Verilog register data types are as rigid as those for nets, but opposite. A reg variable may not be used to connect instances or as the target of a continuous assignment outside of a procedural block. They are assigned values inside of initial and always blocks.

Verilog regs may be either signed or unsigned. Some sample register declarations are shown in Figure 3.6. Signed variables may extend sign bit when used with suitable arithmetic operators. Unsigned regs do not automatically extend any bit, although unsigned variables may be manipulated to preserve a sign bit through replication of the most significant bit.

```
reg A; //Single bit unsigned register variable
reg [15:0] B, C; //Two 16-bit unsigned vectors, 15 is MSB
reg [0:31] D; //A 32-bit unsigned vector, 0 is MSB
reg signed [15:0] E, F; //two 16-bit signed vectors
```

■ **FIGURE 3.6** Sample register declarations

Slices of a vector may be selected and assigned to another variable. To grab the most significant bit of reg B of Figure 3.6 and assign it to a variable called SIGNBIT, the following code could be used:

```
reg SIGNBIT;
always @(B) SIGNBIT = B[15];
```

Similarly, the entire most significant byte could be selected as follows:

```
reg [7:0] HIGHBYTE;
always @(B) HIGHBYTE = B[15:8];
```

An alternative to specifying the high and low bits of a vector slice is to specify a number of bits. This unusual syntax can be useful when using a variable to set the index. In the following examples, both assignments would grab bits 31 through 24:

```
//HIGHBYTE and HB2 will both be assigned bits 23 through 31
always @(DATA_BUS) HIGHBYTE = DATA_BUS[31-:8];
always @(DATA_BUS) HB2 = DATA_BUS[24+:8];
```

The first example says to take eight bits starting with bit 31, not 24 bits from 31 through eight. The second says to take eight bits starting from bit 24. Moving the + sign after the colon, however, means to take 24 bits:

```
//THREEBYTES will be assigned bits 31 down to 8
always @(DATA_BUS) THREEBYTES = DATA_BUS[31:+8];
```

While seemingly obscure, this technique can be useful when stepping although a large variable with an index variable:

```
INDEX = 8;
TARGET = BUS[(INDEX*8 )-:8];
```

As INDEX is decremented, different bytes will be sent to variable TARGET. A construct like this would typically be in a loop. Verilog loops will be covered in Chapter 5.

An odd feature of bit selects is that they are always unsigned, even if the entire vector is selected. This is true even if both the object of the bit select and the target are signed variables.

When a smaller signed variable is assigned to a larger one, the expected result is to extend the sign bit into the larger variable. This will happen if the variable alone is referenced. However, if the variable uses a bit select, even if the select covers the entire vector, the sign bit will not be extended. An example of this is shown in Figure 3.7. In that code, the sign bit of A will be extended into X but not into Y.

```
module SIGNED_ASSIGN (A, X, Y);
  input signed [7:0] A;
  output reg signed [15:0] X, Y;

  always @(A) begin
    X = A; //A will be sign extended to 16 bits
    Y = A[7:0]; //No sign extension
  end

endmodule
```

■ **FIGURE 3.7** Bit selects are always unsigned

With SystemVerilog, it is possible to entirely fill a vector with all ones, zeros, Xs, or Zs. Examples of filling arbitrarily sized variables are shown below in Figure 3.8. This is more useful when the sizes of the variables are set with parameters and are unknown until compile time. Parameters are covered in Chapter 4.

SystemVerilog variables

SystemVerilog introduces several new variable types that have more relaxed rules on where they can be assigned. Some of these new variable types break with standard Verilog in that they only support two values, logic zero and logic one. Two-state variables can never be set to Z or X values.

```
module vectorfill;
  reg [31:0] W;
  reg [15:0] X;
  reg [7:0] Y;
  reg Z;

  always_comb begin
    W = '0; //W will be 32 bits of zeros
    X = '1; //X will be 16 bits of ones
    Y = 'X; //Y will be eight bits of unknown
    Z = 'Z; //Z will be one bit set to high impedance
  end
endmodule
```

■ **FIGURE 3.8** Filling a vector with all 0, 1, X, or Z values

The data types added to SystemVerilog are summarized in Table 3.4. Unlike Verilog integers, which are vaguely defined to be at least 32 bits, these new data types are explicitly defined as shown. The vector types can be any length. All these new data type names are SystemVerilog keywords and thus may no longer be used as variable names.

Table 3.4 Variable types added in SystemVerilog

Name	Function
byte	Two-state signed 8-bit integer
shortint	Two-state signed 16-bit integer
int	Two-state signed 32-bit integer
longint	Two-state signed 64-bit integer
bit	Two-state signed vector
logic	Four-state unsigned vector
shortreal	Two-state signed 32-bit floating point
void	No value, used for functions that do not return a value

Bit and logic types can be used in place of both wires and regs. They are not restricted to being assigned either inside or out of functional blocks. Like wires and regs, they will be one-bit wide if not given a range when declared. Neither logic nor bit data types have resolution functions, so they are unsuitable for use on Tri-state signals, as are all data types other than nets.

Uninitialized two-state variables default to all zeros. Uninitialized four-state variables default to X. Use of two-state variable types can lead to misleading simulation results, as the initial values of a simulation will not necessarily correspond to what the synthesized hardware will do. If proper circuit operation depends on initial conditions, as is frequently the case with state machines, a reset or other method of putting the circuit into a known initial state is still needed despite any simulation data indicating that a design using two-state bit data types automatically starts out at all zeros.

When a four-state variable is connected to a two-state variable, any X and Z values of the four-state variable will be translated to zeros in the two-state variable. This may cause design errors to be missed in verification.

In SystemVerilog, variables also may be declared to be constants by adding the keyword "const" to their declarations. Examples of this are shown below.

```
const int ALPHA = 32;
const reg [15:0] BETA = 14;
const logic [11:0] GAMMA = 1023;
```

Constants are not useful for circuit design as they are assigned their values at initialization time in simulation and not at all in hardware. Their use is thus restricted to nonsynthesizable modules. This is covered further in Chapter 4.

Var variables

In SystemVerilog, variables can be declared to be type "var." This construct is rarely used and may not be supported by all design automation tools. When it is used, its main purpose is to explicitly force an input to be a single-driver net.

Net variables by default can have multiple drivers. This property, along with the resolution functions that allow Tri-state buses, can lead to primary inputs of a device being verified having conflicting values. This is because in the absence of any other declaration, inputs default to nets.

By declaring an input also to be a var, the input is forced to have only a single driver. Connecting a var input to more than one source would be a compilation error. Other signals may also be declared to be type var, although doing so does not add anything to circuit design.

A sample declaration of an input that is also a var variable is shown below.

```
input var A;
```

A variable can be declared to be a var by itself or a var in conjunction with another type. When var is used by itself, the resulting variables will act indistinguishably from logic variables. When used in conjunction with other variable types, use of the keyword var will be without effect for both simulation performance and synthesized circuits. Var can be used with enumerated types (covered in Chapter 4) as well as other built-in Verilog and System-Verilog types.

```
//These three declarations will work identically
var [15:0] A_BUS;
var logic [15:0] A_BUS;
logic [15:0] A_BUS;
```

Arrays

Logic, bit, net, and register variable types can all be made any size, but are one-dimensional. Verilog allows multidimensional arrays of all variable types. They are created by simply adding one or more ranges to the variable declaration.

Two-dimensional arrays can be used to model memories. A memory synthesized from behavioral code would not be a true memory, but would be a register file made from flipflops. Such a device would be an order of magnitude larger and more power hungry than one made with an equivalent-sized memory block. The FIFO example developed in Chapter 7 and implemented in Chapter 12 does use a synthesizable array. For power efficiency, it could be replaced with a dual-port random access memory (RAM).

Some sample array declarations are shown in Figure 3.9. Arrays are typically declared from zero to range minus one, as is done in the first three examples of Figure 3.9, but other styles are legal.

```
module array_decs;

  //a one kilobyte register file
  reg [7:0] MEM [0:1023];

  //an array of 100 16-bit integers
  shortint NUMBERS [0:99];

  //an array of 32 one-bit regs
  reg BIT_ARRAY [0:31];

  //declare only the size, not the range
  logic [15:0] SVMEM [512]; //elements will be 0 to 511

endmodule
```

■ **FIGURE 3.9** Array declarations

SystemVerilog allows a size, rather than a range, to be declared when creating an array. This is shown in the last example of Figure 3.9. When this style is used, the elements will always range from zero to (size − 1).

The third example of Figure 3.9 would create an array of 32 single-bit devices. Such an array is theoretically different from a single vector 32 bits wide, but in hardware the two would be indistinguishable.

Any number of one-dimensional variables may be declared with a single statement, but arrays need to be declared individually. For example, the following line will create five 13-bit regs, but the next line will create one 5 × 8 array ARRAY1 and a single five-bit variable SCALAR1.

```
reg [12:0] A, B, C, D, E; //five 13-bit registers
reg [4:0] SCALAR1, ARRAY1 [0 :7]; //one five-bit reg and one 5 x 8 array
```

Not only may arrays of variables be inferred, but arrays of instances may also be inferred. In the next example, an array of eight Tri-state buffers is inferred:

```
bufif1   BUF_ARRAY[0:7] (Z, A, EN);
```

An array of eight of the above arrays could also be created with a single line of code as follows, assuming the first array is in a module called BUF8:

```
BUF8   BUFFS[7:0] (Z, A, EN);
```

The instance names in these examples, BUF_ARRAY and BUFFS, are arbitrary. Any legal Verilog identifiers may be used.

Z and A can be connected to 64-bit buses without specifying anything further. The 64 enable inputs can all be connected to a single signal or broken out into smaller pieces.

Arrays are not limited to two dimensions. A cubic structure could be created as follows:

```
reg [15:0] CUBE [0:15][0:15];
```

Rather than declare a range for an array, a size may be used. Thus, the following declaration would also create a 16 × 16 × 16 array:

```
reg [15:0] CUBE[16][16];
```

When specifying a size rather than a range, the elements always start at zero and end at (size − 1).

Just as bit selects may be used to take a piece of a vector, word selects may be used to reference one or more elements of an array. In the following example, one word from the cubic memory CUBE is put on a data bus.

```
reg [15:0] DATA_BUS;
DATA_BUS = CUBE[5][6];
```

It is not necessary to select an entire element of an array. In the next example, three bits of one word are taken:

```
reg [2:0] SLICE;
SLICE = CUBE[12:10][9][8];
```

Multiple words of an array can also be selected by using a range in one or more of the fields.

SystemVerilog allows multiple values to be assigned to array elements with a single statement. Thus, a default value may be assigned to all elements of an array, or individual values may be assigned to discrete elements. While synthesizable, some thought should be given to the sort of hardware being inferred through the use of this enhancement.

Examples of multiple element assignments are shown in Figure 3.10. In this SystemVerilog module, the entire array is also used as a port. This is not legal in standard Verilog. In the first assignment, all elements of the array are set to zero. In the second, a hexadecimal value is assigned to element 0, two four-bit inputs are assigned to elements one and two, and a four-bit binary value is assigned to the last element. These are examples of the possible and legal, although not of good style or sense.

```
module array_assign(input CLK, RST,
    input [3:0] DATA1, DATA2,
    output reg [3:0] MEM [4]);

    always_ff @(posedge CLK, negedge RST)
    if (!RST) MEM <= '{default:0};
    else MEM <= '{4'hA, DATA1, DATA2, 4'b1010};
endmodule
```

■ **FIGURE 3.10** Assigning to all elements of an array with single statements

SystemVerilog has also added several synthesizable functions for determining information about arrays. They are shown in Table 3.5. None of these functions manipulate the values of elements in arrays. They only return data about arrays.

Table 3.5 Array functions

Name	Function
$left	Returns the left range limit (MSB) of the array dimension.
$right	Returns the right range limit (LSB) of the array dimension.
$low	Returns the minimum of $left and $right.
$high	Returns the maximum of $left and $right.
$increment	Returns 1 if $left is greater than or equal to $right. Returns −1 otherwise.
$size	Returns the number of elements in a dimension of an array.
$dimensions	Returns the number of dimensions of an array.

An example of using these functions is shown in Figure 3.11. In that example, a four-dimensional array MEM is created and then one slice of it is used as an argument to the functions, except for $dimensions, which operates on the whole array. The values

```
module array_funcs;
  reg [3:0] MEM[4] [5] [6];
  int A, B, C, D, E, F, G;

  assign A = $left(MEM[2][2][2]);
  assign B = $right(MEM[2][2][2]);
  assign C = $low(MEM[2][2][2]);
  assign D = $high(MEM[2][2][2]);
  assign E = $increment(MEM[2][2][2]);
  assign F = $size(MEM[2][2][2]);
  assign G = $dimensions(MEM);

  initial $display("A = %d, B = %d, C = %d, D = %d, E = %d, F = %d, G = %d",
  A, B, C, D, E, F, G);
endmodule


/*$ A = 3, B = 0, C = 0, D = 3, E = 1, F = 4, G = 4*/
```

■ **FIGURE 3.11** Array functions

```
/*Array with output connected to a bidirectional bus.*/

module MEMCELL(input OE, WS, input [7:0] ADDR,
  inout wire [15:0] DATA);

  /*One 16 bit reg variable(DATA_REG), one 16x256 bit array (MEMORY).
  Only MEMORY will infer flipflops. DATA_REG will be a combinational
  decode of the array, but it is still a Verilog reg variable.*/
  reg [15:0] DATA_REG, MEMORY[0 : 255];

  //Registers cannot be connected to an inout port
  assign DATA = OE ? DATA_REG : 16'bz;

  //Select one 16 bit element of the memory array
  always_comb DATA_REG = MEMORY[ADDR];

  //Write to memory on rising edge of write strobe (WS)
  always_ff @(posedge WS) MEMORY[ADDR] <= DATA;
endmodule
```

■ **FIGURE 3.12** Memory cell connected to bidirectional bus

returned by each function are shown as a comment below the source code.

Arrays are most commonly used to model memory cells. Figure 3.12 shows one implementation of an array used to model a two-dimensional register file, which can be used as a general-purpose memory. In actual hardware design, such a flipflop-based register file would only be used for the smallest memories. They have the advantage of speed, but are power hungry and take a lot of area when compared to static or dynamic RAMs.

Bidirectional buses

Arrays as declared in Figure 3.9 may be used as memory cells, although instantiating a RAM cell may be more power and area efficient. Regardless of what technology the memory cell uses, it may need a bidirectional bus.

Bidirectional ports may not be connected to register variables. Attempts to do so will result in compilation errors. The reason for this is that register variables have no resolution functions, but a bidirectional bus needs to resolve being driven simultaneously to high impedance by one or more sources and valid ones or zeros by another. Wires and other net types can resolve the conflicting values. If a wire is connected to one driver that is in high-impedance mode and other that is logic one, the net will be driven to logic one.

To connect a register variable to a bidirectional port, it is thus necessary to first connect the register to a net variable. In Figure 3.12, the connection is done with the "assign" statement. It uses the conditional operator to select either the output of the register array or 16 bits of high impedance. When synthesized, an array of Tri-state buffers will be instantiated. The conditional operator is examined in detail toward the end of this chapter.

The memory cell design of Figure 3.12 uses an output enable (OE) signal to determine the direction of the bidirectional bus. If it is logic one, the output of the memory array is driven onto the bus. If it is logic zero, the port is set to high impedance and may be driven by an external device.

If an OE is logic one and an external device is simultaneously driving the bus, there will be contention for any bits that are driven to logic one by one source and logic zero by another. In a physical circuit, this would lead to a short circuit and damage to the device. In simulation, contention will show up as unknown values. It is up to the user of the design to ensure that this never happens. Exactly one driver should be active at any time.

In Figure 3.12, a write strobe (WS) is connected to the clock inputs of the flipflops in the array. In practice, it might be preferable to create a write enable (WE) signal and connect a master clock signal to the flipflop clock inputs. The lines below would do that and could be substituted for the always_ff block in Figure 3.12.

```
always_ff @(posedge CLOCK)
  if (WE) MEMORY[ADDR] <= DATA;
  else MEMORY[ADDR] <= MEMORY[ADDR];
```

To use the above code snippet, a CLOCK input would need to be added to the input list and the write enable signal WE would need to be substituted for write strobe WS.

Using a WE signal, the memory contents will only be changed when that signal is active. The memory contents will be preserved otherwise, which is done by feeding the current contents back to the cell. In the design of Figure 3.12, new data are written on every rising edge of the WS signal.

Structures and unions

Related variables may be grouped together into structures and unions. While structures and unions look very similar, they are different in concept and function.

Unions are in effect a method of creating aliases for variables. All members of the union will reference the same bits, whether in simulation or synthesized hardware. To be synthesizable, a union must be "packed" and each element must reference the same number of bits. Packed refers to how variables are stored on the host computer in simulation. When variables are packed, there are no unused bits between variables, even if the variables are smaller than the host computer word. Unpacked is the default, as it allows variables to be stored in successive word locations, eliminating any need for the simulator to track where in each word variables start and stop and generally increasing simulation performance.

In Figure 3.13, the three fields A, B, and C all refer to the same eight bits. When simulated, a change to one will be reflected on all three. This is illustrated in the simulation output shown in Figure 3.13. Assignment is made only to A, but the results are seen equally on B and C.

Variables included in a union may be register types or System-Verilog types. Inclusion of any net type in a union would be an error.

```
module unn;

  union packed {
    reg [7:0] A; //A, B and C all reference the same space
    bit [7:0] B;
    byte C;
  } AB;

  initial begin
    $monitor("%d: A = %h, B = %h, C = %h", $time, AB.A, AB.B, AB.C);
    AB.A = 8'hAA; //B and C are never directly assigned any values
    #1 AB.A = 8'h55;
  end

endmodule


# 0: A = aa, B = aa, C = aa
# 1: A = 55, B = 55, C = 55
```

■ **FIGURE 3.13** The union will reference the same space for all three names

For the fields of a union to reference the same space, all fields must be sized identically. The code of Figure 3.14 is defective. D is a 16-bit integer and thus cannot share space with A, B, and C. This code would fail in both simulation and synthesis.

```
module unn;

  union packed {
    reg [7:0] A; //A, B and C all reference the same space
    bit [7:0] B;
    byte C;
    shortint D; //Error: 16 bit value
  } AB;
endmodule
```

■ **FIGURE 3.14** All fields of a packed union must be the same size

Properly sized packed unions are synthesizable. The code of Figure 3.15 would result in eight flipflops being inferred.

```
module unn2(input CLK, input [7:0] A, output wire [7:0] Z);

  union packed {
    reg [7:0] X;
    bit [7:0] Y;
  } XY;

  assign Z = XY.Y;
  always_ff @(posedge CLK) XY.X <= A;

endmodule
```

■ **FIGURE 3.15** Synthesis of module unn2 would yield eight flipflops

Note that the scoping operator, introduced in Chapter 2, is used to reference fields of the union. This technique is also used for other hierarchical references such as named blocks, subroutines, and interfaces as well as modules.

The fields of structures are independent variables. They are grouped together for convenience and clarity rather than functionality. They do not all need to be the same size. In Figure 3.16, the signals of a typical computer bus are grouped into a structure called BUS. BUS itself has a structure instantiated in it, a two-signal control bus consisting of a WS and an OE.

```
module strs;

  typedef struct {bit WS, OE;} CON;

  struct {bit [15:0] ADDR, DATA;
    CON CBUS;} BUS;

  reg [15:0] MEM [0:1023];

  always_ff @(posedge BUS.CBUS.WS) MEM[BUS.ADDR] <= BUS.DATA;

endmodule
```

■ **FIGURE 3.16** A hierarchical system of structures

As with unions, individual signals in the structure may be referenced with hierarchical names and scoping operators. Thus to reference the write strobe, not only is the name of the signal, WS, used, but also the name of the top-level structure, BUS, and the name of the instance of the control bus, CBUS. The name of the type of the control bus, CON, is not used in the hierarchical path. This is analogous to referencing instances of primitives rather than the primitive type, as was done in Chapter 2.

Figure 3.16 also introduces user-defined types and their keyword, typedef. Creating a type is necessary if a structure is going to be instantiated, although a structure need not be instantiated to be used. BUS is an example of creating a structure and using it without instantiation. If a type is created, any number of instances of that type can then be used. BUS, however, can only be the sole representative of that structure, since it is not a defined type.

While it is legal to use structures as ports, there is a limitation: all fields of a structure used in a port list must be the same direction. Because interfaces (covered in Chapter 6) provide similar grouping capability without this limitation and have additional capabilities, interfaces are more commonly used as ports.

An example of using a structure in a port list is shown in Figure 3.17. All fields included in the structure are inputs. The

```
module strs(IBUS, DATA);
  typedef struct {bit [15:0] ADDR;
    bit WS, OE;} BUS;
  input BUS IBUS;
  inout [15:0] DATA;
  reg [15:0] MEM [0:1023], DATA_REG;

  assign DATA = IBUS.OE ? DATA_REG : 16'hz;
  always @(IBUS.ADDR, IBUS.WS) DATA_REG = MEM[IBUS.ADDR];
  always_ff @(posedge IBUS.WS) MEM[IBUS.ADDR] <= DATA;

endmodule
```

■ **FIGURE 3.17** Using a structure in a port list

bidirectional DATA field must be removed from the structure and declared separately. Order of declarations is inflexible in that the typedef is needed before the input statement referencing the type.

Structures can be left at the default of unpacked or explicitly made packed. Unlike unions, structures are not required to be packed to be synthesizable.

All fields of a structure may be assigned values with a single statement and without explicitly referencing each field. In Figure 3.18, the concatenation operator (curly braces) is used to string together a series of values that are assigned to successive elements of structure BUS. Multiple assignments to successive fields as shown in Figure 3.18 will only work if both structures are modified to be packed.

```
module strs;

  typedef struct packed {bit WS, OE;} CON;

  struct packed {bit [15:0] ADDR, DATA;
    CON CBUS;} BUS;

  assign BUS = {16'h4, 16'haa, 1'b0, 1'b1};

endmodule
```

■ **FIGURE 3.18** Referencing all fields of a structure

Typedef also can be used to create new names for built-in variable types. In the following examples, SystemVerilog number types are given aliases to match C language names.

```
typedef shortint short;
typedef longint longlong;
typedef real double;
typedef shortreal float;
```

OPERATORS

Assignment operators

Verilog has two assignment operators: blocking and nonblocking. The rule of thumb is to use the blocking assignment operator with code that will infer only combinational logic and the nonblocking in functional blocks that will infer flipflops.

The terms blocking and nonblocking refer to the order in which expressions are evaluated. When using the blocking assignment operator, the assignments to operands on the left side of the operator are completed before evaluation of expressions to the right of subsequent statements in the block is begun. Evaluation of those subsequent expressions is thus "blocked" until assignments are made to previous variables. In the case of nonblocking assignments, all expressions to the right of the assignment operators are evaluated when the block is entered and only after all have been evaluated are assignments to the left side operands performed. Order of evaluation has significant implications for both simulation performance and synthesized hardware.

In Figure 3.19, the blocking assignment operator is used in the recommended manner. Because evaluation of each statement will proceed in sequence, the updated values for the earlier operations will be used for subsequent ones.

```
module blocking(input [3:0] A, B, C, D,
   output reg [4:0] X, Y, output reg [5:0] Z);

   always_comb begin
     X = A + B;
     Y = C + D;
     Z = X + Y;
   end

endmodule
```

■ **FIGURE 3.19** The blocking assignment operator

In Figure 3.20, a sequential circuit is inferred using the blocking assignment operator. At first glance, this code looks like it should produce a shift register, as shown in Figure 3.21. However, synthesis of this module would produce two registers in parallel, as shown in Figure 3.22, or only one register and the other optimized out entirely. Simulation of the code shown in Figure 3.20 would also show the two registers operating in parallel.

```
module sr2(input CLOCK, DATAIN, output reg FF2);
  reg FF1;
  always_ff@(posedge CLOCK) begin
    FF1 = DATAIN;
    FF2 = FF1;
  end
endmodule
```

■ **FIGURE 3.20** Faulty use of the blocking assignment operator in a sequential block

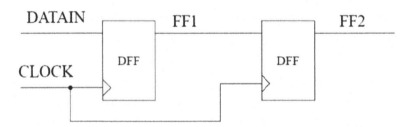

■ **FIGURE 3.21** The intended result, a shift register

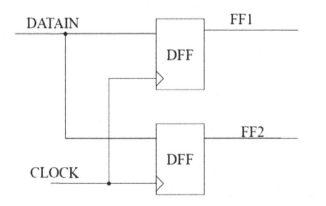

■ **FIGURE 3.22** The actual result, parallel registers

The issue is order of evaluation. The block is triggered by the rising edge of the clock. When that occurs, DATAIN is evaluated and assigned to FF1. Once that assignment has been completed, and without any advancement of the simulation time, the second statement is evaluated. Since, after evaluating the first assignment, FF1 has been set equal to DATAIN, FF2 is then also set equal to DATAIN. The result is two flipflops operating in parallel rather than in sequence.

The easiest way to get a shift register is to replace the blocking assignment operators with nonblocking ones as shown in Figure 3.23. This code will produce a shift register in both simulation and synthesis.

```
module sr2(input CLOCK, DATAIN, output reg FF2);
  reg FF1;
  always_ff@(posedge CLOCK) begin
    FF1 <= DATAIN;
    FF2 <= FF1;
  end
endmodule
```

■ **FIGURE 3.23** Code for a shift register using the nonblocking assignment operator

The lesson that might be drawn from this example is to always use the nonblocking assignment operator. Doing so, however, would cause different problems. In the example of Figure 3.19, if the assignment operators were to be replaced by nonblocking operators, Z would use stale data. The block would only be evaluated when A, B, C, or D changed. Since the nonblocking assignment operator would cause all three expressions to be evaluated before any assignments are made, Z would be assigned the sum of X and Y using their values before they were updated as a result of whatever event caused the block to be evaluated.

Use of stale data would only be a simulation effect. The synthesizer would generate the same gates whichever operator was used, leading to simulation-synthesis mismatches.

Another way to make the code of Figure 3.19 simulate correctly with nonblocking assignment operators would be to use a sensitivity list (Verilog 95 style) and adding X and Y to that list. This would be a poor solution, however, as it would cause the block to be reevaluated from the beginning when X and Y change, giving slower simulation performance.

The code of Figure 3.20 can be made to work as a shift register by changing the order of the assignments. Verilog worked that way before the nonblocking assignment operator was added to the language, but ensuring that the order of assignments is always in reverse order to data flow is awkward and error-prone. Putting each register assignment into a separate always block could also be made to work, but the code would be hard to understand and maintain. Even keeping all registers in separate blocks does not always prevent unexpected operation, as will be demonstrated further in Chapter 8. The recommended technique is to use nonblocking assignment operators for sequential blocks and blocking assignment operators for combinational blocks.

The code of Figure 3.20 can be made to appear to work in simulation via addition of explicit delays, but as already noted, delays are ignored by synthesizers. This would again result in simulation-synthesis mismatches and defective hardware.

In general, a block should use either blocking or nonblocking assignment operators, not some of each. However, most tools do support some mixing of the two in a single block. The schematic diagram shown in Figure 3.24 has both combinational and

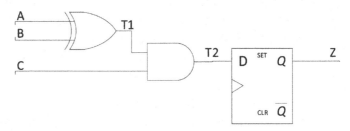

■ **FIGURE 3.24** Digital circuit with combinational and sequential components

sequential components. There are several ways to encode it in Verilog, including mixing assignment operators in a single block.

One option is to write two blocks: one sequential and one combinational, as shown in Figure 3.25. An equally valid representation is shown in Figure 3.26, where the combinational function is merged into a single sequential assignment. Most experienced designers tend to use the method shown in Figure 3.26 until the right-hand side expression gets overly complicated, but overly complicated is not a precise definition. Just when an expression becomes complicated enough to warrant breaking it out into a separate functional block or even a subroutine varies from engineer to engineer.

```
module CS1(input A, B, C, CLOCK, output reg Z);
  reg T1, T2;
  always_comb begin
    T1 = A ^ B;
    T2 = C & T1;
  end

  always_ff@(posedge CLOCK)
    Z <= T2;
endmodule
```

■ **FIGURE 3.25** Separate combinational and sequential blocks

```
module CS2(input A, B, C, CLOCK, output reg Z);
  always_ff @(posedge CLOCK) Z <= C & (A ^ B);
endmodule
```

■ **FIGURE 3.26** One sequential block

A third method is to make combinational assignments and sequential assignments all in one block, using the blocking assignment operator for the combinational parts and nonblocking for sequential, as shown in Figure 3.27.

```
module CS3(input A, B, C, CLOCK, output reg Z);
  reg T1, T2;
  always_ff@(posedge CLOCK) begin
    T1 = A ^ B;
    T2 = C & T1;
    Z <= T2;
  end
endmodule
```

■ **FIGURE 3.27** Legal mixing of assignment operators in a single functional block

While synthesizers do allow assignment operator mixing as shown in Figure 3.27, using both assignment operators to target a single variable is unequivocally illegal for hardware design. Simulators allow this, as they can change the scheduling of assignments to a variable on the fly, but a circuit cannot modify itself as data values change. Thus, the code of Figure 3.28 could be simulated but would be rejected by a synthesizer.

```
module NIX(A, B, C, Z);
  input A, B, C;
  output reg Z;
  always @(A or B or C)
    if (A) Z = A & B;
    else Z <= A | B;
endmodule
```

■ **FIGURE 3.28** Illegal mixing of assignment operators

Equality operators

Equality operators are used to compare operands. They generate a single-bit output based on the comparison. The difference between logical and case operators is that logical operators output an X if either operand is X or Z while case operators compare all bits, including those having X and Z values.

Accordingly, circuit descriptions are limited to using the logical operators, as comparisons to X or Z would be meaningless in hardware. Other than the wildcard version (covered below) the case operators are not synthesizable but they are useful in test fixtures, where detecting X and Z values can be of interest. The symbols for the equality operators are shown in Table 3.6 and their operations in Tables 3.7 and 3.8. Some examples of equality operator usage are shown in Figure 3.29. The inequality operators yield the inverse in each case. Note that the inverse of X is still X.

It is interesting to note that replacing the construct VAR1 = (VAR2 == VAR3) as used in Figure 3.22 with an "if" construct will yield different results in simulation but not in synthesis. For example,

```
if (A == C) U = 1'b1;
else U = 1'b0;
```

would result in U being assigned the value 1'b0 for the values of A and C used in Figure 3.22. The "if" construct used above does not have the option of assigning U an indeterminate value.

Table 3.6 Equality operators

Symbol	Operation
==	Logical equality
!=	Logical inequality
===	Case equality
!==	Case inequality

Table 3.7 Logical equality operator

==	0	1	X	Z
0	1	0	X	X
1	0	1	X	X
X	X	X	X	X
Z	X	X	X	X

Table 3.8 Case equality operator

===	0	1	X	Z
0	1	0	0	0
1	0	1	0	0
X	0	0	1	0
Z	0	0	0	1

```
module equality;

  reg [2:0] A, B, C, D;
  integer E, F;
  reg U, V, W, X, Y, Z;

  initial begin
    A = 3'b11z;
    B = 3'b01x;
    C = 3'b11z;
    D = 3'b011;
    E = 3;
    F = 7;

    //Equality operator, no comparison to Z
    U = (A == C); //U = 1'bx
    //Case equality, does compare to Z
    V = (A === C); //V = 1'b1
    //Equality, no comparison to X, either
    W = (B == D); //W = 1'bx
    //Case equality, resolves X is not Z
    X = (B === D); //X = 1'b0
    //Compares integer and vector values
    Y = (D == E); //Y = 1'b1
    //Smaller vector is zero-extended
    Z = (D == F); //Z = 1'b0
  end
endmodule
```

■ **FIGURE 3.29** Equality operator examples

Since the equality operator does not evaluate X or Z values, the test for equality between A and C fails, even though they are identical. Since assigning X is not an option, the only remaining possibility is to assign U a value of 1'b0.

While the logical equality operators will not compare X or Z values, they will resolve variable comparisons if they can do so from the determinate bits. In Figure 3.30, there is a bit that is zero in A where it is one in B. That is enough to determine that A is not equal to B. Both Y and Z will be set to 1'b1. There is no ambiguity in either case.

```
module equality2;

  reg [4:0] A, B;
  reg Y, Z;

  initial begin
    A = 5'b1010x;
    B = 5'b00101;
    Y = (A != B); //A is not equal to B even without the last bit
    Z = (A !== B); //Logical and case inequality work the same in this case
  end
endmodule
```

■ **FIGURE 3.30** Resolving equality/inequality from determinate bits alone

SystemVerilog has expanded the case equality/inequality to add
wildcard bits to comparisons. These expansions are synthesizable. These extensions are used to mask out specified bits when
operands are compared. Examples of synthesizable uses of wildcards with case equality and inequality operators are shown in
Figure 3.31.

```
module wildcards(input [2:0] A, B, output logic [2:0] C, D);
  always_comb begin
    C = (A ==? 3'b10?); //C will be true for A = 100 and 101;
    D = (B !=? 3'b?00); //D will be true whenever the two LSBs of B are not 00
  end
endmodule
```

■ **FIGURE 3.31** Synthesizable wildcard case equality and inequality operators

In the simplistic example of Figure 3.31, the wildcard bits will be
optimized out of the circuit entirely in synthesis, as there are no
occasions when the least significant bit of A or the most significant bit of B is needed.

Logical operators

Verilog supports several variations of Boolean operators. The
logical operators are used to return a true/false condition.

They always result in a single-bit output, no matter how many bits wide the input operands are. They are typically used in multiway branching structures. Multiway branching is covered in Chapter 5.

Logical operations only support AND, OR, and NOT functions. There is no logical XOR function, although XOR is supported in the bitwise Boolean operators, covered next.

The logical operators are shown in Table 3.9. Some typical examples of their application are shown below.

```
always @(posedge CLK)
//if not enabled, hold state
  if (!ENABLE) DATA <= DATA;
//otherwise load new data
  else DATA <= NEWDATA;

always_comb
if (A && B) NEXT_STATE = FETCH;
if (STATE == DECODE || STATE == IDLE) BUS = 32'bz;
```

Table 3.9 Logical operators

Symbol	Operation
!	NOT
&&	AND
||	OR

In the above examples, the variables being tested are true if each has at least one bit that is logic one. Otherwise they are false. The logical operators do not perform bitwise operations on the operands. Thus in Figure 3.32, C would be set to 1, even though there is not a single-bit position where the two operands ANDed together would produce a logic one.

```
module logicals;
  reg [7:0] A, B;
  reg C;
  always_comb begin
    A = 8'b10101010; //A has some bits set, so it is true
    B = 8'b01010101; //Same for B
    if (A && B) C = 1; //A is true, B is true, so C gets set
    else C = 0;
  end
endmodule
```

■ **FIGURE 3.32** Logical operation, AND of 1 and 1 yielding 1

Ambiguous signals such as A in Figure 3.33 can produce surprising results. In that example, A is tested for both true and false. It fails in both cases, so both D and E will be set to zero.

```
module logicals2;
  reg [7:0] A;
  reg D, E;
  always_comb begin
    A = 8'b0000000x;
    if (A)  D = 1;
    else D = 0;
    if (!A)  E = 1;
    else E = 0;
  end
endmodule
```

■ **FIGURE 3.33** A is neither true nor false. Both D and E will be 0

This is a simulation phenomenon. It has no implications for circuits designed with logical operators, as X is a software concept.

Bitwise operators

The bitwise operators are the ones that perform standard Boolean operations. The bitwise operators are summarized in Table 3.10. Bitwise inversion can be combined with XOR to form an XNOR function but NAND and NOR operators cannot be similarly formed. A &~B would be A ANDed with not B, not A NAND B.

A ~&B would just be a syntax error. A bitwise NAND can be accomplished through use of parenthesis: ~(A&B) would be a bitwise NAND. Bitwise OR operators work in an analogous manner.

Table 3.10 Bitwise operators

Symbol	Operation
~	Bitwise inversion
&	Bitwise AND
|	Bitwise OR
^	Bitwise XOR
~^ or ^~	Bitwise XNOR

Bitwise operators perform the indicated operation with corresponding bits in two operands. If one operand is shorter, it will be extended with zeros to match the length of the longer operand.

In bitwise operations, a Z input is treated as X. Bitwise operators never output high-impedance state.

Bitwise operators will resolve unknown and high-impedance input values according to Tables 3.11–3.14. Some examples of resolution with bitwise operators are shown in Figure 3.34.

Table 3.11 Bitwise AND resolution

&	0	1	X	Z
0	0	0	0	0
1	0	1	X	X
X	0	X	X	X
Z	0	X	X	X

Table 3.12 Bitwise OR resolution

|	0	1	X	Z
0	0	1	0	0
1	1	1	1	1
X	X	1	X	X
Z	X	1	X	X

Table 3.13 Bitwise XOR resolution

^	0	1	X	Z
0	0	1	0	0
1	1	0	X	X
X	X	X	X	X
Z	X	X	X	X

Table 3.14 Bitwise inversion resolution

~	Output
0	1
1	0
X	X
Z	X

```verilog
module bitwise;
  reg [4:0] A, X, RATS;
  reg [3:0] B, V, W, Z;
  reg [2:0] C, D, Y;

  initial begin
    A = 5'b10101;
    B = 4'b010x;
    C = 3'b011;
    D = 3'b11z;

    V = ~B; //Inverse of X is X
    W = A | B; //B will be zero-extended by one bit
    X = A & B; //Indeterminate bit will cause indeterminate output
    Y = C ^ D; //High impedance bit will act like unknown
    Z = B ^~ C; //XNOR, zero extend C
    $strobe("V = %b, W = %b, X = %b, Y = %b, Z = %b", V, W, Y, Y, Z);
  end
endmodule

/*
# V = 101x, W = 0101, X = 10x, Y = 10x, Z = 100x
*/
```

■ **FIGURE 3.34** Bitwise operators

For single-bit operands, bitwise and logical operators work identically.

Reduction operators

Reduction operators also perform Boolean functions but are unary, which is to say that they only operate on a single operand. They produce a single-bit output regardless of the size of the input operand. Their symbols are shown in Table 3.15, and some examples of their usage are shown in Figure 3.35. These operators

Table 3.15 Reduction operators

Symbol	Operation
&	AND reduction
~&	NAND reduction
|	OR reduction
~|	NOR reduction
^	XOR reduction
~^ or ^~	XNOR reduction

```
module reduction;
  reg [7:0] A, B, C;
  reg W, X, Y, Z;
  initial begin
    A = 8'hFF; //All ones
    B = 0; //All zeros
    C = 8'b01100000; //Even number of set bits
    W = &A; //AND reduction: detect max. value
    X = |B; //OR reduction: detect all zeros
    Y = ^C; //XOR reduction: positive parity
    Z = ~^C; //XNOR reduction: negative parity
    $strobe("W = %b, X = %b, Y = %b, Z = %b", W, X, Y, Z);
  end
endmodule

/*
# W = 1, X = 0, Y = 0, Z = 1
*/
```

■ **FIGURE 3.35** Reduction operators

are commonly used for parity generators and to detect a maximal value of a variable. OR reduction can equally easily be used to detect that a variable either is or is not all zeros, but there are other, more intuitive ways of doing that.

Arithmetic operators

Verilog supports the fundamental arithmetic operations of addition, subtraction, multiplication, and division as well as modulus and exponential. All are synthesizable except exponential, which is not currently supported although that is subject to change with new versions of synthesizers. An example of each is shown in Figure 3.36.

```
module ARITH(input [7:0] A, B, output reg [8:0] SUM, DIF,
   output reg [15:0] PROD, output reg [7:0] QUO, REM,
   output reg [1023:0] EXP);

   always_comb begin
     SUM = A + B; //Addition
     DIF = A - B; //Subtraction
     PROD = A * B; //Multiplication
     QUO = A / B; //Division
     REM = A % B; //Modulus
     EXP = A ** B; //Exponential
   end

endmodule
```

■ **FIGURE 3.36** Arithmetic operators

Synthesis tools have progressed so much that not only are addition and subtraction optimized to the point that handcrafting operators at the gate level are a waste of time, so are multiplication and division. In previous generations of the tools, division was limited to powers of two for the divisor, or right shift, but any integer can now be used. Given the proper constraints, modern synthesizers will pick the optimal algorithm for the operation, scale it for the operand size, and do it faster than human designers can typically do. It is now extraordinarily difficult to beat the machine, in either design time or quality of results.

These operators can work with signed or unsigned numbers and any data type. One sometimes unexpected feature of the modulus operator is that the sign of the result always takes the sign of the first operand. Thus,

−5 % −2

would yield −1. The two negatives do not cancel each other out with this operator, although they would with multiplication or division. While not entirely intuitive, it is sensible. A negative number can only have a negative remainder, as repeated subtractions of either a positive number or a negative one will always leave either no remainder or a negative one.

In Figure 3.36, the result of the exponential operation is assigned to a kilobit-long register. Even this register would be nowhere near big enough to store the maximum results of an eight-bit number raised to an eight-bit number, which may account for the lack of synthesis support for the exponential operator.

When operating with integers, it must be understood that the result will be an integer. Disregard of this constraint may lead to surprising results. In the example of Figure 3.33, integer division will result in a half period of zero and the clock never toggling. The code can easily be fixed by changing PERIOD to a floating point type such as real. A parameter (parameters are covered in Chapter 4) with a value of 1.0 will also work, but a parameter set to 1 will have the same problem as the code shown in Figure 3.37.

Auto increment and auto decrement

SystemVerilog adds auto increment and decrement operators, which are not available in standard Verilog. They can be pre or post, for a total of four different options. Increment and

```
module divzero;
  const int PERIOD = 1;
  reg CLK;
  initial begin
    CLK = 1'b0;
    //Integer division: 1 / 2 = 0
    #(PERIOD / 2) CLK = 1'b1;
    #(PERIOD / 2) CLK = 1'b0;
    #(PERIOD / 2) CLK = 1'b1;
    #(PERIOD / 2) CLK = 1'b0;
  end
endmodule
```

■ **FIGURE 3.37** Integer division truncates

decrement operators are typically used in loop controls, as shown in Figure 3.38.

```
for (I = 0; I < LIMIT; I++) //post increment
for (I = LIMIT; I > 0; I--) //post decrement
for (I = 0; I < LIMIT; ++I) //pre increment
for (I = LIMIT; I >= 0; --I) //pre decrement
```

■ **FIGURE 3.38** Auto increment and decrement operators used in loop controls

These operators are all blocking assignment operators, which can lead to illegal mixing of assignment types. In Figure 3.39, an attempt is made to create a counter. This piece of code would simulate correctly, but it is not a viable hardware description. This design would fail in synthesis because CNT is properly assigned using the nonblocking assignment operator in the RST clause, but the increment would be blocking, resulting in a scheduling conflict, and a synthesis error. To make the counter synthesizable, CNT++ should be replaced with CNT <= CNT+1.

```
always @(posedge CLK, negedge RST)
  if (!RST) CNT <= 'b0; //non-blocking assignment
  else CNT++; //blocking assignment
```

■ **FIGURE 3.39** Illegal mixing of blocking and nonblocking assignments

Relational operators

Relational operators, shown in Table 3.16, are used to compare two variables and return a single-bit value reflecting the relationship between them. They are similar to the logical equality operators with one difference: they will not resolve the relationship between operands whenever there is an X or a Z in an operand. A single X or Z bit in either operand at any bit position is enough to make the result of the operation ambiguous in simulation. As with the equality operators, ambiguous results are only a simulation effect. Physical circuits can never take on an X value, so the discrepancy in handling indeterminate bits has no effect on hardware design.

Table 3.16 Relational operators

Symbol	Operation
>	Greater than
<	Less than
>=	Greater than or equal to
<=	Less than or equal to

Figure 3.40 has some examples of applying relational operators. There is no relational operator analogous to the case equality/inequality operators. Relational operators all fail to resolve X and Z values. This difference in behavior is illustrated in Figure 3.41.

The less than or equal operator symbol is identical to the nonblocking assignment operator. Compilers determine the meaning from the context. There is no place where both would be meaningful. This is illustrated in Figure 3.42, where the first use of <= is as a relational operator and the remaining ones are nonblocking assignment operators.

Shift operators

Arithmetic logic units frequently need to shift or rotate data, and these types of operations are facilitated with shift operators.

```
module relational;

  integer A, B;
  reg [2 : 0] C;
  reg [1:0] D;
  reg W, X, Y, Z;

  initial begin
    A = 5;
    B = 17;
    C = 3'b11x;
    D = 2'b0z;

    W = (A < B); //True: W = 1'b1
    X = (B <= A); //False: X = 1'b0
    Y = (C < B); //Will not resolve: Y = 1'bx;
    Z = (D >= A); //Will not resolve: Z = 1'bx;
  end
endmodule
```

■ **FIGURE 3.40** Relational operator examples

```
module resolve;

  reg [3:0] A, B;
  reg X, Y, Z;

  initial begin
    A = 4'b1000;
    B = 4'b000x;

    //Relational will return unknown
    X = (A > B); //X = 1'bx
    //Logical equality will resolve A is not equal to B
    Y = (A == B); //Y = 1'b0
    //Case equality will also resolve A is not equal to B
    Z = (A === B); //Z = 1'b0
  end
endmodule
```

■ **FIGURE 3.41** Difference in resolution between relational and equality operators

```
always_ff @(posedge CLK)
  for (int I = 0; I <= MAX; I++)
    if (ENABLE) OUT <= OP1 + OP2;
    else OUT <= OUT;
```

■ **FIGURE 3.42** $<=$ can mean a relational or an assignment operator

Verilog has four of these operators covering right and left shifts, arithmetic and simple. Shifting left and right is the same as multiplying and dividing by powers of two, although any remainders are discarded. The shift operators are shown in Table 3.17.

Table 3.17 Shift operators

Symbol	Operation
$\ll$	Shift left
$\gg$	Shift right
$\lll$	Arithmetic shift left
$\ggg$	Arithmetic shift right

The difference between arithmetic shift left and simple shift left is that arithmetic extends the sign bit when used with signed operands. Arithmetic shift right works exactly the same as a simple shift right under all conditions.

The shift operators use two operands or constants, one to provide the data and the other the shift distance. Examples are shown in Figure 3.43. If the sign bit is not extended, all the shift operators fill the shifted-out bits with zeros. The bits shifted out are lost. There is no provision for directly saving them in another variable, although doing so is not hard to arrange.

The code of Figure 3.43 will yield the following results:

A = 11111111, W = 00111111, X = 11111100, Y = 11111111, Z = 11111100

Y does not change at all and would not change with any shift distance. It will always sign extend to remain all ones. The others

```
module shift;
  reg signed [7:0] A, W, X, Y, Z;

  initial begin
    A = 8'hFF; //Initialize to 11111111
    W = A >> 2; //Shift right by 2 bits (divide by 4)
    X = A << 2; //Shift left by 2 bits (multiply by 4)
    Y = A >>> 2; //Arithmetic shift right, sign extend
    Z = A <<< 2; //Arithmetic shift left
  end
endmodule
```

■ **FIGURE 3.43** Shift operations

all fill with zeros. If, however, Y was declared to be an unsigned variable, it would not have a sign bit to extend, so it would work like a simple shift.

A variable-length shifter can be created by simply replacing the constant in the above example with another variable, as shown below:

NEW = OLD >> DISTANCE;

Shift operations are sometimes useful, but wrapping the shift bits around, as is done in a barrel shifter, is more common. There is no single operator that can do this, but a barrel shifter can be designed by adding a small amount of code to a variable shifter. One implementation of a synthesizable, scalable barrel shifter is shown in Figure 3.44. It uses parameters, which are covered in Chapter 4.

```
module barrel(A, SHIFT, OUT);
  parameter DIST = 2;
  parameter WIDTH = 2**DIST;
  input [WIDTH - 1 : 0] A;
  input [DIST - 1 : 0] SHIFT;
  output reg [WIDTH - 1 : 0] OUT;

  always @(A, SHIFT) OUT = {A, A} >> SHIFT;
endmodule
```

■ **FIGURE 3.44** Barrel shifter using a shift operator

It also uses an exponential operator, which might seem to make the design nonsynthesizable, but that is not the case. It remains synthesizable because the exponential operator is only used for variable sizing, which is fixed at compile time. This is true for both simulation and synthesis.

The barrel shifter of Figure 3.44 uses the concatenation operator, covered next, to duplicate variable A.

Streaming, also called packing and unpacking, is a new System-Verilog enhancement of shifting. Packing is when a streaming operator is used on the right-hand side of an assignment operator. It streams bits starting from the most significant bit position of the source to the least bit of the destination, effectively reversing the bit order. Unpacking is when the streaming operator is used on the left-hand side of an assignment operator. Then a bit stream starting with the least significant bit of the source is assigned to bits starting with the least significant of the destination.

Examples of streaming are shown in Figure 3.45. Use of curly braces as shown is required.

```
module streaming;
  logic [15:0] A, B, C;
  logic [7:0] D, E;
  initial
    A = 16'h1234;
  always_comb begin
    B = {<<{A}};
    {>>{C}} = A;
    D = {<<{ A[7:0]}};
    {>>{E}} = A;
    $strobe("A = %h, B = %h, C = %h, D = %h, E = %h", A, B, C, D, E);
  end
endmodule

//# A = 1234, B = 2c48, C = 1234, D = 2c, E = 12
```

■ **FIGURE 3.45** Streaming operation

Concatenation operator

Two or more variables can be strung together to operate as one with the concatenation operator. It uses curly braces, which may explain why Verilog uses "begin…end" where the C programming language on which the syntax of Verilog is based uses curly braces: all the keys on a standard keyboard are used in Verilog. There was nothing else left.

A simple application of concatenation is shown in Figure 3.46, where the sum and carry out bits are concatenated to allow a single line of code to feed both variables.

```
module adder(A, B, CIN, SUM, COUT);
  input A, B, CIN;
  output reg SUM, COUT;

  always @(A or B or CIN) {COUT, SUM} = A + B + CIN;
endmodule
```

■ **FIGURE 3.46** Adder using concatenation

One caveat on the concatenation operator is that it always yields an unsigned quantity, even if all concatenated variables are signed.

A common application of the concatenation operator is to group together multiple variables to form a single return value from a Verilog function. Subroutines, including functions, will be covered in Chapter 6.

Replication operator

The replication operator is similar to the concatenation operator but it adds the ability to make multiple copies of all or part of a variable.

How the replication operator works when the replication constant is zero has changed over the years. In early versions of Verilog, this was an undefined condition. Most design automation tools inserted one bit of logic zero when that was encountered. Since this is obviously incorrect, the 2001 language revision finessed the problem by declaring a zero replication constant to be illegal. Finally in 2005, the language was revised to eliminate any term with a zero replication constant. This last behavior was carried over to SystemVerilog.

The main use of the replication operator used to be for extending sign bits. This was made unnecessary when Verilog was expanded to include signed variables, although it still can be used that way when working with unsigned types.

The replication constant must indeed be a constant. Use of a variable in that spot would cause a compilation error. Use of parameters, however, blurs the distinction between a variable and a constant enough to allow a replication constant of zero in some instances of a design. More information on parameters will be forthcoming in Chapter 4.

In the example of Figure 3.47, X will be set to 10101010, Y will be set to 00001010, and Z will be set to 11111010. The results for Y may be different if compiled with pre-Verilog 2005 versions of tools.

```
module replication;
  reg [3:0] A;
  reg [7:0] X, Y, Z;
  initial begin
    A = 4'b1010;

    X = {2{A}}; //make two copies of A concatenated together
    Y = {{0{A[3]}}, A}; //zero replication constant
    Z = {{4{A[3]}}, A}; //Extend A's MSB four more bits
  end
endmodule
```

■ **FIGURE 3.47** Replication examples

Conditional operator

The conditional operator is unusual in that it can be used to infer multiplexors or Tri-state drivers. Examples of each are shown in Figure 3.48. Both are synthesizable. In the first case, a two to one multiplexor would be created. The second would cause a synthesizer to select a Tri-state driver, assuming Tri-state buses are supported by the target library.

```
module conditional(input A, B, SEL, EN,
output reg OUT);
  reg C;

  always_comb begin
    C = SEL ? A : B;
    OUT = EN ? C : 1'bz;
  end
endmodule
```

■ **FIGURE 3.48** Conditional operators creating a multiplexor and a Tri-state driver

Use of the conditional operator for creating multiplexors is generally desirable because such constructs inherently resolve ambiguous data. What this means in practice is that if the select line is unknown, the output bits will only be unknown, or ambiguous, for bits that are in conflict on the data inputs. This is illustrated in Figure 3.49. As is always the case with unknown outputs, ambiguous bits are a simulation phenomenon. In synthesized hardware, they will be either logic one or logic zero, although it is impossible to know which.

SystemVerilog combined assignment operators

SystemVerilog adds the ability to combine arithmetic, logical, and shift operators with blocking assignment operators. These operators are shown in Table 3.18. There are not any combined nonblocking operators.

```
module conditional2;
  reg [2:0] A, B, C;
  reg SEL;

  always_comb C = SEL ? A : B;

  initial begin
    #0 A = 3'b010; B = 3'b101; SEL = 1'b0;
    #1 SEL = 1'b1;
    #1 SEL = 1'bx;
    #1 B = 3'b011;
  end

initial $monitor($time, "A = %b, B = %b, C = %b, SEL = %b", A, B, C, SEL);

endmodule

/*
# Time 0: A = 010, B = 101, C = 101, SEL = 0
# Time 1: A = 010, B = 101, C = 010, SEL = 1
# Time 2: A = 010, B = 101, C = xxx, SEL = x
# Time 3: A = 010, B = 011, C = 01x, SEL = x
*/
```

■ **FIGURE 3.49** Multibit multiplexor resolving common bits

Table 3.18 SystemVerilog combined assignment operators

Symbol	Operation
+=	Add right side to left side, assign to left side
-=	Subtract right side from left side, assign to left
*=	Multiply left side by right side, assign to left side
/=	Divide left side by right side, assign to left side
%=	Divide left side by right side, assign remainder left
&=	Bitwise AND left and right sides, assign to left side
|=	Bitwise OR left and right sides, assign to left side
^=	Bitwise XOR left and right sides, assign to left side
≪=	Shift left side left by number of bits on right, assign to left side
≫=	Shift left side right by number of bits on right, assign to left side
≪≪=	Arithmetic shift left by number of bits on right, assign to left side
≫≫=	Arithmetic shift right by number of bits on right, assign to left side

While these operators save a small amount of typing, they do not add any new functionality. Some examples of their use are shown in Figure 3.50. In the examples of Figure 3.49, X would end with a value of 4'b1111, Y with 4'b0010, and Z with 4'b1111.

```
module combined;
  reg signed [3:0] A, B, C, X, Y, Z;

  initial begin
    A = 4'b1010;
    B = 4'b0001;
    C = 4'd2;
    X = 4'h5;
    Y = 4'b1;
    Z = 4'b1111;
    X += A; //Add X + A, assign to X
    Y <<= B; //Shift Y left by B bits
    Z >>>= C; //Arithmetic shift right Z by C bits
  end
endmodule
```

■ **FIGURE 3.50** Combined assignment operators

Operator precedence

Instead of simply evaluating expressions from left to right, Verilog uses precedence as shown in Table 3.19. Depending on operator precedence can produce obscure code. In the example shown below, precedence will cause the expression to be evaluated from right to left. It is unlikely that the casual reader would correctly interpret it.

X = ^ A & B >> 3;

By judicious use of parentheses, the meaning can be clarified:

X = ^ (A & (B>>3));

The two expressions are identical, but the meaning of the first is obscure. Maintenance of a design containing such structures would be a fraught exercise. The second one is clear and easy to maintain.

Table 3.19 Operator precedence, from highest to lowest

Operator	Meaning
() []	Parenthesis, Bit Select
+ - ! ~ & ~& | ~| ^ ~^ ^~ ++ --	Unary, Autoincrement, Autodecrement
**	Exponential
* / %	Multiplication, Division, Modulus
+ -	Addition, Subtraction
≪ ≫ ⋘ ⋙	Shift
≪= ≫=	Relational
== != === !==	Equality, Inequality
&	Bitwise AND
^ ~^ ^~	Bitwise XOR, XNOR
|	Bitwise OR
&&	Logical AND
||	Logical OR
?:	Conditional
= += -= *= /= %= &= ^= |= ≪ = ≫= ⋘ = ⋙ = <=	Assignment
{} {{}}	Concatenation, Replication

This is not to say that parentheses should be used with wanton abandon. In the following expression, over use of parenthesis may cause a synthesizer to produce suboptimal circuitry. By forcing a specific order of evaluation, the synthesizer's flexibility in adjusting the circuit to signal arrival times will be limited.

SUM <= (A*B) + ((C*D) + (E+F) + G));

The same expression can be written without parenthesis, maintaining identical logic and allowing optimization for timing as appropriate:

SUM <= A*B + C*D + E + F + G;

Operators at the same level of precedence are evaluated left to right.

■ SUMMARY

Modern design methodology encourages a top-down flow, where the overall goals of the design are first clarified and the implementation details filled in later.

Initial blocks have no meaning in circuit design but are essential for verification. Several varieties of always blocks are used for circuit descriptions. Use of continuous assignments is legal but discouraged.

Several different variable types are useful for circuit design. Traditional Verilog users have gone entire careers using nothing but reg and wire operands, but SystemVerilog has added some new and useful types. There are also some nonsynthesizable floating point types.

All variable types can be made into arrays. SystemVerilog has also added structures and unions for gathering together related operands.

This chapter has presented the operators available in Verilog and SystemVerilog. Almost all are synthesizable and may be used in circuit descriptions, but a few are not and their use must be confined to nonsynthesizable verification modules.

## Chapter 4

Behavioral coding part II: defines, parameters, enumerated types, and packages

Code readability and maintainability can frequently be enhanced through the use of mnemonic names and global text substitutions. Verilog supports several different methods for these techniques, as well as providing for design reuse and scalability.

GLOBAL DEFINITIONS

The compiler directive `define is used for string substitution. Once a macro has been defined, it becomes available throughout the design hierarchy.

As a compiler directive, `define does not follow Verilog language conventions. It does not need to be inside of any module. Unlike a standard line of Verilog code, it is terminated by a new line, not by a semicolon. If a defined macro does conclude in a semicolon, the semicolon becomes part of the text substitution string.

The global nature of defined macros is both a strength and a weakness. Because it has the power to change values in any module at any level of hierarchy, `define must be used with care. A common discipline is to have a single definitions file that all designers on a project reference but that only one person maintains. In this way, values that should be common across the entire scope of the design will remain so and unexpected changes to values in design files avoided. In Figure 4.1, a define macro is created and used inside of a verification module. While syntactically correct, defining this global parameter more centrally may be a better strategy, as PERIOD as defined in this module will be global, possibly overwriting other uses of PERIOD.

```
module define1(output reg CLK);
  `define PERIOD 10
  initial begin
    CLK = 1'b0;
    forever #`PERIOD CLK = ~CLK;
  end
endmodule
```

■ **FIGURE 4.1** Defining a value

Defined macros do not necessarily have a value at all. The act of defining one can provide all the utility needed. This is a technique used in conditional compilation. If a macro is defined, regardless of any value associated with the macro, a section of code will be compiled. Conditional compilation also uses `ifdef, `ifndef, `elsif, and `endif directives. The available define directives are shown in Figure 4.2. In that example, the `elsif clause and the `ifndef clause will work identically. There are two ways of accomplishing the same selection.

Only one macro can be defined as per use of `define. Attempting to define multiple macros with one use of `define as shown below

```
module ifdirectives;
  `define ABC //Define but do not give a value to string ABC
  `ifdef ABC //execute the following code if ABC has been defined
    //if defined code goes here
  `elsif //execute the following code if the previous `ifdef was false
    //alternate code goes here
  `endif //end of conditional execution block
  `ifndef ABC //execute the following code if ABC has not been defined
    //if not defined code goes here
  `endif //needed to terminate `ifndef block
  `undef ABC //ABC is no longer defined
endmodule
```

■ **FIGURE 4.2** Define directives

would result in syntax errors whenever STRING1 was referenced. STRING2 would remain undefined.

```
//STRING1 would be defined. STRING2 would not be
`define STRING1  ABC STRING2 DEF
//This will result in a syntax error, as STRING1 will include 3 strings and 2 spaces
reg `STRING1;
```

To define both STRING1 and STRING2, the following syntax would work.

```
`define STRING1 ABC
`define STRING2 DEF
```

In Figure 4.3, if the line `define FAULT is included in the code, the design unit MYDES_FAULTED will be compiled. If it is

```
module define2;
  `define FAULT
  `ifdef FAULT
    MYDES_FAULTED UUT(A, B, CLK, Q);
  `else
    MYDES UUT(A, B, CLK, Q);
  `endif
endmodule
```

■ **FIGURE 4.3** Conditional compilation

deleted or commented out, a version without faults will be selected. This technique can be used in verification to determine if a self-checking test fixture is capable of finding inserted flaws. It can also be used to select between different implementations of a design, such as the original source code or a synthesized gate-level representation. Note that FAULT has no value. The only consideration is whether or not it has been defined.

Negative logic also is supported with defined macros: if a macro is not defined, an action that would not otherwise be necessary can be taken. This can be useful in a design environment that is supposed to have a global definitions file but that file may not be implemented at the time when some simulations are run. In Figure 4.4, the period will be defined locally if it has not already been defined, but the local definition will not be made if the period has already been defined elsewhere. That example also uses an initial condition when declaring CLK to be a port and a variable of type reg. While the initial condition assignment would work in simulation, it would be ignored by synthesizers if this technique were used in a hardware design module. Although syntactically similar to implicit continuous assignments, initial assignments to register variables are entirely different in that they are transitory and for simulation purposes only. In contrast, not only is an implicit continuous assignment synthesizable, but the value so assigned will be permanent.

```
module define3(output reg CLK = 1'b1);
  `ifndef PERIOD
    `define PERIOD 10
  `endif

    initial forever #`PERIOD CLK = ~CLK;
endmodule
```

■ **FIGURE 4.4** Checking if a macro has not already been defined

The contents of a sample definitions file are shown in Figure 4.5. Note that it does not need to be in any sort of a module structure. A series of macros is all the file contains.

```
`define TRUE 1'b1
`define FALSE 1'b0
`define PERIOD 10
```

■ **FIGURE 4.5** A definitions file

This file may be referenced by compiling it along with the source code files that will use the definitions. Alternatively, it may be explicitly included in source code files by using an `include directive, as shown in Figure 4.6. Note that the quotation marks are required. The `include directive is not limited to definition files. Any piece of Verilog code may be so referenced, which will cause the code to act as if it were typed into the source code. This technique is most often useful with subroutines where the subroutine code is maintained in its own file but is repeatedly used in other design and verification modules. Subroutines are covered in Chapter 6.

If the code of Figure 4.6 were to be synthesized, only the hardware inferred in the final "always" block would result. The clock generator, including the initial assignment of CLK to logic one, is nonsynthesizable and would be ignored.

```
module use_include(input A, B, output reg C);
  `include "definitions.v"
  //TRUE, FALSE and PERIOD are defined in definitions.v
  reg CLK = `TRUE; //initial assignments are non-synthesizable
  initial forever CLK = #`PERIOD ~CLK;
  always @(posedge CLK)
    if (A == `FALSE) C <= A & B;
    else C <= A | B;
endmodule
```

■ **FIGURE 4.6** Including a definitions file

A defined macro can be removed from operation by undefining it via the `undef directive. This is shown below.

`undef PERIOD

While some recommend undefining a macro before redefining it to a new value, this author finds that an eccentric exercise. The recommendation here is to never undefine or redefine macros. Their strength is their universality, in that they maintain constant values across the design hierarchy. Undefining and redefining a macro makes the values used dependent on the order of compilation, which is likely to lead to unintended consequences and errors.

Defined macros can take arguments, although this feature is not universally supported. The following example only multiplies together two variables, but the expressions that use the arguments can be arbitrarily complex.

```
`define MUL(M, N) M*N   //MUL macro forms the product of two arguments
```

The macro could then be used in a module to form the product of two variables, as shown below.

```
input [7:0] A , B;
wire [15:0] W;
assign W = `MUL(A, B); //Use the MUL macro to multiply A and B
```

Because defined macros always end with white space, in traditional Verilog there is no way to use arguments to a macro to concatenate together strings such that the resultant string does not have any white space. SystemVerilog extends the define macro capability by adding a new delimiter that can be used for this purpose. This new delimiter is two-backtick characters. In the example below, use of the macro catname would create variables with the arguments to the macro concatenated together. The result would be the declarations of five-bit register variables alphabravo and charliedelta.

```
`define catname(A,B) reg [4:0] A``B;   //two backticks, not a quotation mark
`catname(alpha, bravo)  //note absence of terminating semicolons
`catname(charlie, delta)
```

This technique could be used to create a series of variables with different index suffixes. However, use of an array for this purpose would be more common. Arrays are covered in Chapter 3.

PARAMETERS

Parameters are used to create scalable, reusable modules. The default values for parameters may be overridden to create new implementations of old designs as well as to create multiple instances of a design with different sizes or other characteristics.

Parameters differ from defined macros in several ways. A parameter is only available in the module in which it is declared, unlike a defined macro, which is global in scope. Parameters may be changed on an instance-by-instance basis, giving unique values to each instance but leaving all other instances unchanged. Each parameter must have a default value, unlike a defined macro, which can exist without a value at all.

A scalable counter using a parameter is shown in Figure 4.7. If the parameter is not overridden, the counter will be four bits.

```
module scalable_counter(CLK, RST, CNT);
 parameter SIZE = 4;
  input CLK, RST;
  output [SIZE - 1:0] CNT;
  reg [SIZE - 1:0] CNT;

  always @(posedge CLK or negedge RST)
    if (!RST) CNT <= 0;
    else CNT <= CNT + 1;
endmodule
```

■ **FIGURE 4.7** Parameterized (scalable) counter

Taking advantage of Verilog 2001 and SystemVerilog extensions, the same counter can be rewritten as shown in Figure 4.8. The two implementations are functionally identical.

```
module scalable_counter #(parameter SIZE = 4)
(input CLK, RST, output reg [SIZE - 1 : 0] CNT);

  always_ff @(posedge CLK, negedge RST)
    if (!RST) CNT <= 'b0;
    else CNT <= CNT + 1;
endmodule
```

■ **FIGURE 4.8** Scalable counter using Verilog 2001 and SystemVerilog enhancements

```
module scalable_counter #(parameter SIZE = 4, MAX_CNT = 11)
(input CLK, RST, output reg [SIZE - 1 : 0] CNT);

  always_ff @(posedge CLK, negedge RST)
    if (!RST) CNT <= 'b0;
    else
      if (CNT == MAX_CNT) CNT <= 0;
      else CNT <= CNT + 1;
endmodule
```

■ **FIGURE 4.9** Scalable, programmable counter

Parameters can be referenced throughout the module, not just when establishing size. In Figure 4.9, the point at which the counter rolls over is set with a second parameter.

Unlike variables, parameters are fixed at compile time. This is true for both simulation and synthesis. This is the reason the example of Figure 3.44 remains synthesizable despite the use of a nonsynthesizable operator. The exponential operator there only operates on a parameter, not a variable, so the final value is determined when the module is compiled. This value does not change during operation.

In these examples, the value of each parameter is an integer. The language allows greater flexibility than that. Any type, including strings, may be assigned to parameters, although the utility in circuit design of anything other than an integer is limited at best.

The following parameter declarations are all legal, albeit not useful for hardware description.

```
parameter mystring = "hidiho";
parameter PATH = "/usr/synopsys/syn/libraries";
parameter FLOAT1 = 5.12, FLOAT2 = 45.456;
parameter Pi = 3.14159, e =2.71828;
```

Multiple parameters may be declared with a single use of "parameter." Multiple parameter declarations are shown in the last two examples above. Attempts to do something similar with `define macros would result in compiler errors rather than two macros per line.

OVERRIDING DEFAULT VALUES

The power of parameters is in their ability to be changed on an instance-by-instance basis. In Figure 4.10, two instances of the scalable counter are made. One will be six bits wide, the other eight. No instance of the default four-bit design will be generated.

```
module counters(input CLK, RST,
  output wire [5:0] CNT1, output wire [7:0] CNT2);

  scalable_counter #(.SIZE(6), .MAXCNT(31)) C1(CLK, RST, CNT1);
  scalable_counter #(.SIZE(8), .MAXCNT(255)) C2(CLK, RST, CNT2);
endmodule
```

■ **FIGURE 4.10** Overriding the default value of parameters

In this example, both parameters in each instance are overridden. There is no requirement to do so. Thus, the following line would override parameter MAXCNT only, leaving the size at the default value of four bits.

```
scalable_counter #(.MAXCNT(13)) C3(CLK, RST, CNT3);
```

Figure 4.10 uses "named parameter redefinition," a feature added to Verilog in the 2001 revision to the language standard. Earlier ways of overriding parameter values remain available.

One of such method is "parameter value assignment." It does not explicitly say which parameter is being overridden but otherwise looks similar to named parameter redefinition. The design of Figure 4.10 is rewritten in Figure 4.11 to use this older method.

```
module counters(input CLK, RST,
 output wire [5:0] CNT1, output wire [7:0] CNT2);
 scalable_counter #(6, 31) C1(CLK, RST, CNT1);
 scalable_counter #(8, 255) C2(CLK, RST, CNT2);
endmodule
```

■ **FIGURE 4.11** Overriding default values via value rather than named assignment

Besides the obvious lack of name association with each parameter, this second method has the disadvantage of not being able to skip a leading parameter. If only MAXCNT needs to be changed in an instance of scalable_counter, using value assignments, a value for SIZE would still need to be specified. However, lagging parameters that use the default do not need to be repeated. Thus, the following line would change the size of a counter but leave the terminal value unchanged.

scalable_counter #(5) C3(CLK, RST, CNT3);

The last method of parameter redefinition is with "defparam." Unlike the other methods of changing a parameter value, defparam has unlimited scoping. Rather than being limited to operating on instantiations in the current design, defparam can change any parameter in any module anyplace in the design hierarchy.

Use of defparam to change one parameter in one instance is shown in Figure 4.12. When used as shown, defparam works just like the other methods: it changes a parameter in an instance in the current design.

In Figure 4.13, a more dangerous use of defparam is illustrated. Rather than operating on a parameter in a module instantiated in the current design, a parameter in a different module is changed. This parameter, which is in a nonsynthesizable test fixture, is then

```
module counters(input CLK, RST,
 output wire [5:0] CNT1, output wire [7:0] CNT2);

 scalable_counter #(.SIZE(6), .MAXCNT(31)) C1(CLK, RST, CNT1);
 defparam C2.SIZE = 8;
 scalable_counter C2(CLK, RST, CNT2);
endmodule
```

■ **FIGURE 4.12** Changing a parameter with defparam

```
module scale_reg #(WIDTH = 1) (input CLK,
input [WIDTH - 1 : 0] D,
output reg [WIDTH - 1 : 0] Q);
  always @(posedge CLK) Q <= D;
  defparam tb_scale_reg.SIZE = 16;
endmodule

module tb_scale_reg;
  parameter SIZE = 4;
  wire [SIZE - 1 : 0] Q;
  reg [SIZE - 1 : 0] D;
  reg CLK = 1'b1;
  initial forever #1 CLK = ~CLK;
  initial begin
    D = 15;
    #2 D = 123;
  end
  scale_reg #(.WIDTH(SIZE)) SR(CLK, D, Q);
endmodule
```

■ **FIGURE 4.13** Risky use of defparam to change a value in a higher-level module

pushed down to the current design. Because the change is made in a module that is part of the simulation hierarchy but not part of the design hierarchy, simulation/synthesis mismatch is likely.

When designs and design teams were small, allowing such anarchistic changes were a feature of the language, much loved by engineers who chafed under the restrictions of strongly typed languages. With multibillion transistor designs now being implemented in Verilog, the freewheeling nature of defparam is often

```
module counters(input CLK, RST,
 output wire [5:0] CNT1, output wire [7:0] CNT2);

 defparam C1.SIZE = 5;
 scalable_counter C1(CLK, RST, CNT1);
 defparam C2.SIZE = 8;
 scalable_counter C2(CLK, RST, CNT2);
 //oops: another override of C2 parameter
 defparam C2.SIZE = 16;
 scalable_counter C3(CLK, RST, CNT3);
endmodule
```

■ **FIGURE 4.14** Last one wins: C2 will be set to 16 wide; C3 will be left at its default value

considered more of a liability, a source of risk, and has gone out of favor.

Another risk associated with defparam is its property of having the last call win. In Figure 4.14, two changes are made to one parameter and none to another, rather than one to each. This is presumably a typographical error, but it is perfectly legal syntax and could result in a defective circuit.

Because hierarchical paths to any parameter from any module are legal, an infinite variety of errors may be introduced maliciously or carelessly through uncontrolled use of defparam. For this reason, its use is banned in many design environments, although it remains supported by current simulation and synthesis tools.

LOCAL PARAMETERS

Local parameters, keyword localparam, have the apparently odd property of not being able to have their values changed. While this might seem to defeat the purpose of having them, they do have a use in circuit design. It is to provide mnemonic names for values. Local parameters were added to Verilog in the 2001 language revision and may be unavailable in some older tools.

A typical use for local parameters is in state variables. While it would be highly unusual to want to override the value of a state

```
module localp(input CLK, RST, output reg [3:0] STATE);
   localparam IDLE = 4'b0001;
   localparam FETCH = 4'b0010;
   localparam DECODE = 4'b0100;
   localparam EXECUTE = 4'b1000;

   always_ff @(posedge CLK, negedge RST)
     if (!RST) STATE <= IDLE;
     else begin
       case (STATE)
         IDLE: STATE <= FETCH;
         FETCH: STATE <= DECODE;
         DECODE: STATE <= EXECUTE;
         EXECUTE: STATE <= FETCH;
       endcase
     end
endmodule
```

■ **FIGURE 4.15** Localparam example

in a state machine, giving meaningful names to states can make a design easier to understand. An example of using local parameters is shown in Figure 4.15. That example makes use of a "case" construct, which is a type of multiway branching. Multiway branching is covered in Chapter 5. The example also uses one-hot encoding. That is not required. As with regular parameters, any value type may be used with localparams.

Enumerated types, covered below, also provide a method of creating meaningful mnemonic names while adding error checking into the variable type, although at the cost of increased complexity and more usage restrictions. Either may be used to give intuitive names to constants. Enumerated types tend to be preferred for complex designs done by large design and verification teams, whereas the simpler local parameters are often preferred for small designs done by an individual or small group.

Enumerated types are a SystemVerilog construct unavailable in standard Verilog.

SPECIFY PARAMETERS

Specify parameters, or specparams, are the last of the parameter types. Their use is restricted to specify blocks, which is where performance characteristics for cell models are set. They will be covered in Chapter 11.

ENUMERATED TYPES

Enumerated types specify a set of values that can be assigned to a variable. Only values specified in the set are legal. Any attempt to assign a value outside of the value set to an enumerated type variable would result in an error. This strong typing represents a complete reversal of the traditional freewheeling assignment rules of standard Verilog, where anything can be assigned to any variable type and mismatched or overflow bits will simply be dropped.

In the following line, enumerated type variable STATE can only be assigned one of three specified values.

```
enum {FETCH, DECODE, EXEC} STATE;
```

Enumerated type variables default to integer size, but no matter how many bits an enumerated type variable is, any assignment of a value other than the ones specified in the declaration would be illegal.

The values in the list do have default values. In the above example, FETCH would be equal to zero, DECODE one, and EXEC two. The default values can be changed with explicit value assignments. In the following line, FETCH is explicitly set to one. With that as a starting point, DECODE will be set to two and EXEC to three.

```
enum {FETCH = 1, DECODE, EXEC} STATE;
```

However, it would still be illegal to directly assign STATE a value of three. Only FETCH, DECODE, and EXEC would work, even though EXEC is equal to three. This is illustrated in Figure 4.16.

```
module etypes;
  enum {FETCH = 1, DECODE, EXEC} STATE;
  initial begin
    STATE = FETCH;
    #1 STATE = DECODE;
    #1 STATE = 3; //This will produce an error
  end
endmodule
```

■ **FIGURE 4.16** Illegal assignment to enumerated type variable STATE

Mixing explicit assignments with automatic incrementing of enumerated types can lead to more than one type having the same value. This is illegal and prevents the code containing the duplicate assignment from compiling. An example of this illegal condition is shown below.

    enum {ALPHA = 1, BETA, GAMMA = 2} LETTERS; //Error: Both BETA and GAMMA are 2

In enumerated type LETTERS, both BETA and GAMMA would be two. The code would not compile, as values must be unique.

Unlike the standard Verilog "int" type, SystemVerilog integers are explicitly defined to be 32 bits. While they are legal for synthesis, this sizing frequently makes them undesirable for circuit design. Instead, vectors properly sized for each variable are typically used.

Rather than using the default 32-bit integers, enumerated types can be sized to avoid creating any superfluous hardware. The local parameter example of Figure 4.15 could be rewritten to use an enumerated type as shown in Figure 4.17. Use of a typedef is necessary when changing the enumerated type from the default integer. It has the added benefit of allowing any number of variables of the newly created type to be added. Thus, once the STATES type has been created, the following lines would become legal:

    STATES STATE, NEXT_STATE; //declare two variables of type STATES
    //block structure, etc...
    STATE = NEXT_STATE; //assign combinational logic to registers

```
module enumsm(CLK, RST, STATE);
  input CLK, RST;
  typedef enum reg [3:0] {IDLE = 4'b0001, FETCH = 4'b0010,
  DECODE = 4'b0100, EXECUTE = 4'b1000} STATES;
  output STATES STATE;
  always_ff @(posedge CLK, negedge RST)
    if (!RST) STATE <= IDLE;
    else begin
      case (STATE)
        IDLE: STATE <= FETCH;
        FETCH: STATE <= DECODE;
        DECODE: STATE <= EXECUTE;
        EXECUTE: STATE <= FETCH;
      endcase
    end
endmodule
```

■ **FIGURE 4.17** Sized enumerated type state machine

This construct is frequently found in state machines where the combinational next state logic is separated from the state registers. An example of such a state machine is included in Chapter 12.

In Figure 4.17, the STATE variable is four bits wide but only 4 of the possible 16 encodings are enumerated, meaning that only those 4 values may be assigned to it. This restriction is the primary difference between using localparams and enumerated types. With localparams, the anticipated values can be given mnemonic names, but there is no restriction on values that can be assigned to variables. With enumerated types, any attempt to assign a value outside of the specified set will produce an error, which can then be fixed during the verification phase of a design.

This feature may be bypassed via casting. In the event that a designer is absolutely determined to assign a value outside of the enumerated set, that value can be cast as shown in Figure 4.18.

Direct assignment of an enumerated value to a variable as is done in Figure 4.18 will always work, but sometimes optimal circuit design calls for navigating around the values by incrementing and

```
module etypes;
  typedef enum {FETCH = 1, DECODE, EXEC} STATE;
  STATE STATES;
  initial begin
    STATES = FETCH;
    #1 STATES = DECODE;
    #1 STATES = STATE'(3); //Casting to enumerated type
  end
endmodule
```

■ **FIGURE 4.18** Casting an integer to an enumerated type

decrementing the current value. Using normal arithmetic operators can cause errors if not all possible values have been enumerated. SystemVerilog includes the methods shown in Table 4.1 to
avoid such errors. The last method in the table, "name," is not
useful for circuit design but may be useful in verification.

Table 4.1 SystemVerilog enumerated type methods

Method Name	Function
first	returns first value of enumeration
last	returns last value of enumeration
next	returns next from current value
prev	returns previous from current value
num	returns number of elements in type
name	returns string representation of current value

In Figure 4.19, the enumerated type STATE has only three defined
values. If, instead of using the "next" method, variable STATES
was incremented by adding one to its current value, the first two
increments would work, but the third would result in an illegal
assignment. This is because when STATES is equal to EXEC,
adding one to it would result in a value of three, which is not
one of the enumerated types. Methods, by contrast, wrap back to
the beginning so FETCH follows EXEC. This incorrect method

```
module methods(input CLK, RST);
    typedef enum {FETCH, DECODE, EXEC} STATE;
  STATE STATES;
  always_ff @(posedge CLK, negedge RST)
    if (!RST) STATES <= FETCH;
    else STATES <= STATES.next;
endmodule
```

■ **FIGURE 4.19** Using a method to step through enumerated types

```
module methods(input CLK, RST);
    typedef enum {FETCH, DECODE, EXEC} STATE;
  STATE STATES;
  always_ff @(posedge CLK, negedge RST)
    if (!RST) STATES <= FETCH;
    /*Bad idea: will result in STATES incrementing
    to 3 instead of wrapping back to FETCH*/
    else STATES <= STATE'{STATES + 1};
endmodule
```

■ **FIGURE 4.20** Defective attempt to step through an enumerated type

of incrementing an enumerated type is shown in Figure 4.20. It could work if all possible bit patterns of the enumerated type were to be defined, but using methods is still the preferred technique.

Simply adding one to STATES as shown below would not even compile.

STATES <= STATES + 1; //Error: must use casting to add to an enumerated type

If incrementing or decrementing by one is insufficient, the methods can take an argument to move by any integer value. This is shown in Figure 4.21, where incrementing by two, decrementing by three, going to the first enumerated value, and going to the last enumerated value are all used. Using these methods, only the enumerated values will be used. Incrementing from the last value will go to the first and decrementing from the first will go to the last.

```
module methods(input CLK, RST, input [1:0] VAL);
    typedef enum {FETCH, DECODE, EXEC} STATE;
  STATE STATES;
  always_ff @(posedge CLK, negedge RST)
    if (!RST) STATES <= FETCH;
    else begin
        //increment by two
      if (VAL == 0) STATES <= STATES.next(2);
        //decrement by three
      else if (VAL == 1) STATES <= STATES.prev(3);
        //set STATES to the first value, FETCH
      else if (VAL == 2) STATES <= STATES.first;
        //set STATES to the last value, EXEC
      else STATES <= STATES.last;
    end
endmodule
```

■ **FIGURE 4.21** Using enumerated type methods

```
module COUNT(CLK, RST, CNT);
  input CLK, RST;
  typedef enum {ZERO, ONE, TWO, THREE}NUMBERS;
  output NUMBERS CNT;

  always_ff @(posedge CLK, negedge RST)
    if (!RST) CNT <= ZERO;
    else CNT <= CNT.next;
endmodule
```

■ **FIGURE 4.22** Counter using enumerated type

Enumerated types can be ports as well as internal variables. In the example of Figure 4.22, an enumerated type is the output of a counter. When synthesized, the output will be indistinguishable from a two-bit counter created with standard binary operands, but in behavioral simulation the output values will be ZERO, ONE, TWO, and THREE rather than binary values.

The enumerated type is then used as the input to a decoder in Figure 4.23. The input is declared to be a variable of type NUMBERS rather than a two-bit vector.

```
module DECODE(CNT, ONEHOT);
  typedef enum {ZERO, ONE, TWO, THREE}NUMBERS;
  input NUMBERS CNT;
  output logic [3:0] ONEHOT;
  always_comb
    case (CNT)
      ZERO: ONEHOT = 4'b0001;
      ONE: ONEHOT = 4'b0010;
      TWO: ONEHOT = 4'b0100;
      THREE: ONEHOT = 4'b1000;
      default: ONEHOT = 4'bx;
    endcase
endmodule
```

■ **FIGURE 4.23** Enumerated type is used as an input

At the top, the enumerated type is passed between the modules. The variable CNT is again of type NUMBERS. It is an input to module DECODE and an output from module COUNT. At the top level of the design, it is a signal that needs to be declared. Otherwise, variable CNT would default to a single-bit wire, which would be incompatible with the signals in the two modules.

A test fixture for the design hierarchy of Figures 4.22–4.24 is shown in Figure 4.25. The enumerated type signal only goes between modules and is not a primary input or output, so it does not need to be declared in the test fixture. A simulation of the system including the internal variable CNT is shown in Figure 4.26.

```
module TOP(input CLK, RST, output logic [3:0] ONEHOT);
  typedef enum {ZERO, ONE, TWO, THREE}NUMBERS;
  NUMBERS CNT;
  COUNT  C(CLK, RST, CNT);
  DECODE D(CNT, ONEHOT);
endmodule
```

■ **FIGURE 4.24** Enumerated types must be specified between the modules, too

```
module tb_top;
 bit CLK, RST;
 logic [3:0] ONEHOT;
 TOP  UUT(CLK, RST, ONEHOT);
  initial forever #1 CLK = ~CLK;
  initial begin
    RST = 1'b1;
    #2 RST = 1'b0;
    #3 RST = 1'b1;
  end
endmodule
```

■ **FIGURE 4.25** Test fixture for enumerated type circuit design

/tb_top/CLK	0							
/tb_top/RST	1							
/tb_top/UUT/CNT	ONE	ZERO	ONE	TWO	THREE	ZERO	ONE	TWO
/tb_top/ONEHOT	0010	0001	0010	0100	1000	0001	0010	0100

■ **FIGURE 4.26** Simulation of enumerated type design

The test bench uses SystemVerilog variable types bit and logic. Since bit variables default to zeros, the clock signal does not need to be initialized before setting it to toggle every time unit. This feature would present some risk if used in a circuit description, as the actual value at power up may be different. This risk factor only applies to synthesizable modules, not test fixtures. The reset signal is initialized to logic one at time zero, overriding the default assignment of logic zero.

The logic variable ONEHOT could equally well be declared to be a standard Verilog multibit wire. The result would be the same.

To avoid declaring the enumerated type in each module, it could be created in a package. The package would then have to be imported into each module. Packages are discussed later in this chapter. Using a package is preferable to declaring the same type in multiple modules, as then the type only needs to be maintained in one place.

CONSTANTS

Constants can be declared with the SystemVerilog keyword "const." While useful in verification suites, they are so far not useful for synthesis as they may only be assigned a value at initialization. Since initial value assignments are ignored for synthesis, the synthesized hardware ends up with constants that do not have any specific value. In simulation, const declarations are assigned their values at the start of run time, rather than at elaboration like parameters and localparams, which can account for the difficulty in implementing this construct in the synthesizable subset of SystemVerilog. For circuit descriptions, parameters and localparams are used to establish constant values.

Values are declared to be constants in conjunction with another declaration that defines their type. A few examples of constant declarations are shown below.

```
//Eight bits, constant value
const reg [7:0] FORTYFIVE = 45;
//Integer, constant value
const integer FIFTYSEVEN = 57;
//Array of four constants, values 1, 2, 3 and 4
//Depends on having parameters WIDTH and DEPTH already declared
const logic signed [WIDTH - 1 : 0] COE [0 : DEPTH - 1] = {1, 2, 3, 4};
```

For verification modules and other nonsynthesizable code, constants are a convenient way to give meaningful names to numbers. Since their values are lost in synthesis, they should never be used in circuit description code. Parameters, localparams, and enumerated types may all be used in circuit descriptions.

PACKAGES

SystemVerilog allows declarations to be shared across a design hierarchy by use of packages. While much of the same effect can be accomplished through `ifdef and `include directives, packages provide a more elegant and comprehensive solution.

Packages can include localparams, enumerated types, constants, and subroutine definitions. Packages can import statements from other packages and export them to other packages.

While parameters also can be included in packages, a parameter in a package cannot be redefined, which defeats the purpose of having a parameter. In a package, parameters and localparams work identically. This is so because packages are not local to any instance; thus, redefining a parameter on an instance-by-instance basis is not possible when the parameter is in a package.

To reference a component of a package, a new SystemVerilog operator, the scope resolution operator, is used. Its symbol is the double colon, "::." Wildcards may be used with the scope resolution operator to reference multiple components or all components. The definitions file of Figure 4.5 is reworked as a package in Figure 4.27. The package can then be imported into a module compiled with the package and elements of the package referenced as if they were declared locally, as shown in Figure 4.28. In this example, all elements of package globe are imported and then may be used in the module.

```
package globe;
   `define true 1'b1
   `define false 1'b0
   `define period 10
endpackage
```

■ **FIGURE 4.27** Global definitions in a package

```
module useglobe(input A, output logic B);
   import globe::*; //import everything in package globe
   always_comb
     if (A == true) B = false;
     else B = true;
endmodule
```

■ **FIGURE 4.28** Importing all elements of a package

In principle, individual elements of the package may be imported. However, doing so presents a problem when those elements are defined macros. In the example of Figure 4.22, if discrete elements were imported rather than the entire package, the macro substitution would take place in the import line itself. Thus, the line

```
import globe::`true;
```

would be interpreted as "import 1'b1;," which would be a syntax error. The macro `true would remain undefined in the importing module.

Changing the define macros to parameter or localparam declarations avoids this problem. An example of this is shown in Figure 4.29. After they are imported, true and false may be used throughout the module as if they had been declared locally, although they cannot be redefined.

```
package globe;
  parameter true = 1'b1;
  parameter false = 1'b0;
  parameter period = 10;
endpackage

module useglobe(input A, output logic B);
  import globe::true;
  import globe::false;
  always_comb
    if (A == true) B = false;
    else B = true;
endmodule
```

■ **FIGURE 4.29** Importing individual elements of a package

Package elements can also be directly referenced without explicit importation. Thus, the following lines creating a clock generator could be used in a test fixture as long as the package is compiled along with the module.

```
logic CLK = 1'b1;
initial forever #globe::period CLK = ~CLK;
```

Using this style, period must be referenced with the package name and scope resolution operator every time it is used.

The hierarchical example of Figures 4.22–4.24 is shown in Figure 4.30 reworked to use a package. It would work with the

```
package types;
   typedef enum {ZERO, ONE, TWO, THREE}NUMBERS;
endpackage

module COUNT(CLK, RST, CNT);
  input CLK, RST;
  import types::*;
  output NUMBERS CNT;

  always_ff @(posedge CLK, negedge RST)
    if (!RST) CNT <= ZERO;
    else CNT <= CNT.next;
endmodule

module DECODE(CNT, ONEHOT);
  import types::*;
  input NUMBERS CNT;
  output logic [3:0] ONEHOT;
  always_comb
    case (CNT)
      ZERO: ONEHOT = 4'b0001;
      ONE: ONEHOT = 4'b0010;
      TWO: ONEHOT = 4'b0100;
      THREE: ONEHOT = 4'b1000;
      default: ONEHOT = 4'bx;
    endcase
endmodule

module TOP(input CLK, RST, output logic [3:0] ONEHOT);
   import types::*;
   NUMBERS CNT;
   COUNT  C(CLK, RST, CNT);
   DECODE D(CNT, ONEHOT);
endmodule
```

■ **FIGURE 4.30** Hierarchical design with a package

same test fixture shown in Figure 4.25, producing an identical output to that shown in Figure 4.26.

In Figure 4.30, the wildcard operator is used to import all the contents of the package. This technique is efficient, as each of the enumerated types would otherwise need to be individually imported. Replacing the import line with

```
import types::NUMBERS;
```

would be insufficient because the types of type NUMBERS would remain undefined in the modules. Each reference to ZERO, ONE, TWO, and THREE would then produce an error.

For the code shown in Figure 4.30 to work, the package needs to be compiled before any of the modules that reference it.

FILLING A SCALABLE VARIABLE WITH ALL ONES

When a variable's size is fixed, it is easy to set all its bits to any value. However, if a variable's size is a parameter, setting all bits to logic one can be tricky. In standard Verilog, it can be done through use of negative numbers, but SystemVerilog has added some new syntax to make it easy and intuitive. The same SystemVerilog syntax can also be used to set all bits to logic zero, X, or Z.

In Figure 4.31, methods for setting all bits of a scalable register are shown [1]. The first three will work in standard Verilog. In the first one, A is set to negative one, which in two's complement arithmetic means each bit will be set to one. Variable B is set to a bitwise negation of zero, which also results in each bit being set to logic one. The third method uses the replication operator to generate SIZE bits of logic one. The fourth method is a new SystemVerilog option. It too will result in each bit being set to logic one. Note that this is not the same as setting D to `b1. That would result in only the least significant bit being set with all others equal to zero.

```
module ALLONES;
  parameter SIZE = 64;
  reg [SIZE - 1 : 0] A, B, C, D;
  initial begin
    //A, B C and D will all be set to 64 bits of logic one
    A = -1;
    B = ~0;
    C = {SIZE{1'b1}};
    D = '1;
  end
endmodule
```

■ **FIGURE 4.31** Setting all bits of a variable to logic one

To set all bits of D to high impedance or unknown, the following lines would work. Similar syntax could also be used to set all bits to logic zero, but would be unnecessary as simply setting the variable to zero without any size or radix would suffice.

```
D = 'Z;
D = 'X;
```

■ SUMMARY

The techniques presented in this chapter are primarily used to enhance understanding of code through use of mnemonic names. Define macros should be used with a great deal of care, as they can redefine elements of modules anywhere in the design hierarchy. Parameters, localparams, and user-defined types are less susceptible to accidental misuse.

Parameters are a powerful tool for creating flexible, reusable design modules. Their capacity for redefinition on an instance-by-instance basis means that scalable design units may be created and reused repeatedly.

Localparams are only used for mnemonic reference. Their values cannot be changed.

User-defined types share some functionality with localparams but add error checking. Their use is especially appropriate in state machines.

Constants work in simulation but their values are not preserved in synthesis. They should not be used in circuit descriptions.

Packages provide a convenient structure for maintaining and sharing these mnemonic devices throughout a design hierarchy.

REFERENCE

[1] Sutherland Stuart, Mills Don. Verilog and SystemVerilog Gotchas. Springer; 2007.

Behavioral coding part III: loops and branches

Verilog supports two types of multiway branching and five looping constructs. SystemVerilog adds two more types of loops that are unsupported in standard Verilog. Loops are more commonly used in verification than circuit design, but most looping constructs are synthesizable as long as the number of iterations is fixed at compile time.

Multiway branching is fundamental for all high-level circuit descriptions as well as verification.

All these constructs except generate loops are only legal inside of functional blocks. Generate statements can exist only outside of any blocks, although they still must be inside of a module.

Digital Integrated Circuit Design Using Verilog and SystemVerilog   978-0-12-408059-1

LOOPS
While loop

A while loop evaluates an expression and enters the loop if the expression is true. The expression can be arbitrarily complex but must resolve to a true or false state. The loop code will repeat as long as the condition remains true. Once it turns false, the loop will be exited.

Unlike the "do while" construct, a Verilog while loop will not run even once if the condition tests false initially. A practical effect of this is that operands probably should not be initialized inside of a while loop.

The essential structure of a while loop is just the keyword "while" followed by an expression in parenthesis that evaluates to either true or false, as shown in the following line.

```
while (MY_EXP)
```

If the operation to be performed while MY_EXP is true consists of more than one statement, a "begin" and "end" are needed as well, as shown below.

```
while (tempreg) begin
  if (tempreg[0]) count = count + 1;
  tempreg = tempreg >> 1; // Shift right
end
```

In the above example, the code in the loop will run as long as tempreg is true. Just what true means can be a source of confusion. The example of Figure 5.1 was written to clarify what counts as true and what does not. In that example, the test condition is initialized to unknown (all bits X). That variable is then checked to see if the loop should be entered.

Simulation of module tb_whileloop reveals that the loop is never entered. A variable set to unknown is not true. At the end of the simulation, both count and loopcount are still zero. No display statements are run.

The code of Figure 5.1 was then modified to set tempreg initially to 8'b1xxxxxxx, as shown in Figure 5.2. Running that code resulted in the output of Figure 5.3.

```verilog
module tb_whileloop();
   reg [7:0] tempreg;
   reg [3:0] count;
   integer loopcount;
   initial begin
    count = 0;
    tempreg = 8'bx;
    loopcount = 0;
    while (tempreg) begin
         #10 //delay to show running counts
         if (tempreg[0]) count = count + 1;
         tempreg = tempreg >> 1; // Shift right
         loopcount = loopcount + 1;
         $display("tempreg = %b, count = %d,
         loopcount = %d", tempreg, count, loopcount);
     end
     end
endmodule
```

■ **FIGURE 5.1** Test fixture for determining true/false evaluation

```verilog
module tb_whileloop();
   reg [7:0] tempreg;
   reg [3:0] count;
   integer loopcount;
   initial begin
    count = 0;
    tempreg = 8'b1xxxxxxx;
    loopcount = 0;
    while (tempreg) begin
         #10 //delay to show running counts
         if (tempreg[0]) count = count + 1;
         tempreg = tempreg >> 1; // Shift right
         loopcount = loopcount + 1;
         $display("tempreg = %b, count = %d,
         loopcount = %d", tempreg, count, loopcount);
     end
     end
endmodule
```

■ **FIGURE 5.2** Code of Figure 5.1 modified to set one bit of tempreg to logic one

```
# tempreg = 01xxxxxx, count =  0, loopcount =      1
# tempreg = 001xxxxx, count =  0, loopcount =      2
# tempreg = 0001xxxx, count =  0, loopcount =      3
# tempreg = 00001xxx, count =  0, loopcount =      4
# tempreg = 000001xx, count =  0, loopcount =      5
# tempreg = 0000001x, count =  0, loopcount =      6
# tempreg = 00000001, count =  0, loopcount =      7
# tempreg = 00000000, count =  1, loopcount =      8
```

■ **FIGURE 5.3**   Output of tb_whileloop with tempreg initialized to 8'b1xxxxxx

The code was then modified again to initialize tempreg to 8'bxxxxxxx1. When that version was run the output was the line shown below.

    # tempreg = 0xxxxxxx, count = 1, loopcount = 1

The conclusion that can be drawn from these experiments is that true means while at least one bit is identically equal to one. It does not mean while all bits are not equal to 0.

A corollary is that

    while (!MY_EXP)

will enter the loop only while each and every bit of the condition is identically equal to zero. Some examples of values that would test false for (!MY_EXP) are shown below.

    tempreg = 8'b10000000;
    tempreg = 8'xxxxxxxx;
    tempreg = 8'0000000x;

While loops may be synthesizable as long as the number of iterations is fixed at compile time, which is to say that it is not data dependent. However, they are more commonly used in test fixtures than in circuit descriptions. A generic template for this sort of use is shown in Figure 5.4.

```
parameter MAX_OUTER = 1023, MAX_INNER = 255;
integer i = 0, j = 0;
initial begin
  while (i <= MAX_OUTER) begin
    while (j <= MAX_INNER) begin
    //run thousands of tests here
    j = j + 1;
    end
  i= i + 1;
  end
end
```

■ **FIGURE 5.4** Typical usage of a while loop in a test fixture

Do while loop

The essential difference between a while loop and a do while loop is that the latter will always be entered at least once, whereas the former will not be entered at all if the condition initially tests false. Thus, the do while loop allows variables to be initialized in the loop. While unlikely to be useful for circuit design, local variables may be created inside the do while structure, as is done with CMINUS in Figure 5.5. Any attempt to reference CMINUS outside of the loop, including using it as the test variable, would

```
module dowhile;
  initial begin
    int CNT = 0;
    do begin
      //CMINUS is local to the do...while loop
      int CMINUS;
      #1 CNT++;
      $strobe("COUNT = %d", CNT);
      CMINUS = CNT - 1;
      $strobe("CMINUS = %d", CMINUS);
    end
    while (CNT < 10);
  end
endmodule
```

■ **FIGURE 5.5** Do while loop

produce a compilation error. If CNT were replaced with CMINUS in the comparison (CMINUS < 10), the code would not compile.

Running the code shown for 10 time units or more would result in displaying the values 1 through 10 for CNT and 0 through 9 for CMINUS.

Do while loops are a feature of SystemVerilog but are not available in standard Verilog.

An example of a loop that initially tests false along with its output is shown in Figure 5.6. Replacing "FALSE" in the condition with anything else would result in an infinite loop.

```
module dowhile;
  initial begin
    string CONDITION = "FALSE";
    do
      $display ("This will print anyhow.");
    while (CONDITION != "FALSE");
  end
endmodule

# This will print anyhow.
```

■ **FIGURE 5.6** Do while loops always run at least once

For loop

For loops perform the same jobs as while loops, but tend to be the construct most favored in both circuit design and verification. They also tend to be used when no loop is needed at all.

A syntactical example of a for loop is shown in Figure 5.7. The loop, however, is useless, in that the same results could be obtained without it by replacing the loop and its internal assignment with nothing but OUT = A & B.

In Figure 5.7, the three-bit variable I is used for loop control. Even if the loop were not useless, that variable would not result in any additional hardware being created. There would be no

three-bit register in any hardware synthesized from the loop. The
loop control simply directs the compiler on how much hardware
to make: four gates, in this case.

```
module forloop(input [3:0] A, B,
output reg [3:0] OUT);
   reg [2:0] I;
   always @(A or B) begin
    for (I = 0; I <= 3; I = I + 1)
      OUT[I] = A[I] & B[I];
   end
endmodule
```

■ **FIGURE 5.7** A syntactically correct but useless for loop

The for loop in Figure 5.7 may look like it is mixing the blocking
and nonblocking assignment operator, but it is not. It is using the
blocking assignment operator and a relational operator.

A design that will count the number of bits set in a register using
a for loop is shown in Figure 5.8. After a value is loaded, the loop
will step through each bit of the target data. Synthesizers will

```
module forloop2 #(WIDTH = 8) (input CLOCK, RESET, LOAD,
input [WIDTH - 1:0] DATA,
output reg [WIDTH - 1 : 0] COUNT);
reg [WIDTH - 1 : 0] INDEX, TEMP;
always @(posedge CLOCK or negedge RESET) begin
  if (!RESET) COUNT <= 0;
  else begin
    if (LOAD) TEMP <= DATA;
    else begin
      for (INDEX = 0; INDEX < WIDTH; INDEX = INDEX + 1) begin
        COUNT <= COUNT + TEMP[0];
        TEMP <= TEMP >> 1;
      end
    end
  end
end
endmodule
```

■ **FIGURE 5.8** Counting ones with a for loop

"unroll" loops and generate as much hardware as needed to fulfill the indicated assignments for each iteration. A surprisingly large amount of hardware can be created this way.

The code of Figure 5.8 can be modified to end early if there are no more ones to be found, as shown in Figure 5.9. Instead of iterating through the loop a fixed number of times, the loop will terminate as soon as the OR reduction of "temp" yields logic zero. While this modification could make simulation of the design run faster, it would also make it nonsynthesizable because the number of iterations through the loop is now data-dependent. To be synthesizable, the number of iterations through a loop must be fixed at compile time, which is not the case in the code of Figure 5.9.

```verilog
module forloop2 #(WIDTH = 8) (input CLOCK, RESET, LOAD,
input [WIDTH - 1:0] DATA,
output reg [WIDTH - 1 : 0] COUNT);
reg [WIDTH - 1 : 0] INDEX, TEMP;
always @(posedge CLOCK or negedge RESET) begin
  if (!RESET) COUNT <= 0;
  else begin
    if (LOAD) TEMP <= DATA;
    else begin
      for (INDEX = 0; |TEMP; INDEX = INDEX + 1) begin
        COUNT <= COUNT + TEMP[0];
        TEMP <= TEMP >> 1;
      end
    end
  end
end
endmodule
```

■ **FIGURE 5.9** Nonsynthesizable for loop

The index registers in Figures 5.8 and 5.9 are made the same size as the data registers. This would seem to be inefficient, as it only needs to be the $\log_2$ of that size, or a ceiling function of the log. Making it exactly as big as needed would actually add inefficiency, though. Verilog does not have built-in log functions, so one would have to be created for the purpose. Even with such a function, running it would cost some time. Since the index does not translate into any additional hardware, making it bigger than necessary has no penalty. Many designers use integers for indices, since the size of the variable that holds the index has no effect on the synthesized hardware and putting in an integer variable without calculating the optimal size is easy.

SystemVerilog has added a ceiling $\log_2$ function, $clog2. It may be used in synthesizable code to create optimally sized variables in scalable designs. Figure 5.10 has the code of Figure 5.8 modified to use the $clog2 function to size the loop control variable INDEX. If synthesized, the two would produce identical hardware. Note that INDEX needs to be sized to $clog(WIDTH) : 0, not $clog(WIDTH) − 1 : 0. With the smaller size, INDEX could never reach eight and an infinite loop would be created. This would never terminate in simulation and would be a synthesis error.

```
module forloop2 #(WIDTH = 8) (input CLOCK, RESET, LOAD,
input [WIDTH - 1:0] DATA,
output reg [WIDTH - 1 : 0] COUNT);
reg [WIDTH - 1 : 0] TEMP;
reg [$clog2(WIDTH) : 0] INDEX;
always @(posedge CLOCK or negedge RESET) begin
  if (!RESET) COUNT <= 0;
  else begin
    if (LOAD) TEMP <= DATA;
    else begin
      for (INDEX = 0; INDEX < WIDTH; INDEX = INDEX + 1) begin
        COUNT <= COUNT + TEMP[0];
        TEMP <= TEMP >> 1;
      end
    end
  end
end
endmodule
```

■ **FIGURE 5.10** Sizing loop control variable with $clog2

For loops may be useful in circuit design but are subject to over-use by inexperienced designers. In Figure 5.11, a for loop is used to create a scalable shift register. While not incorrect, the loop is not needed. The second module in Figure 5.11 implements the same design without a loop.

In the first module of Figure 5.11, the index variable I is declared as it is assigned. This is legal in SystemVerilog but not in standard Verilog.

```
/*Shift register using an unnecessary loop*/
module shiftreg #(DEPTH = 8) (input CLK, DATA_IN,
  output DATA_OUT);
  reg [DEPTH - 1 : 0] SHIFT;
  assign DATA_OUT = SHIFT[DEPTH - 1];
  always_ff @(posedge CLK) begin
    SHIFT[0] <= DATA_IN;
    for (int I = 0; I < DEPTH; I++)
      SHIFT[I + 1] <= SHIFT[I];
  end
endmodule

/*Shift register without a loop. Both will
synthesize to the same hardware but this one
will simulate faster.*/
module SR2#(DEPTH = 8) (input CLK, DATA_IN,
  output reg DATA_OUT);
  reg [DEPTH - 2 : 0] SHIFT;
  always_ff @(posedge CLK)
    {DATA_OUT, SHIFT} <= {SHIFT, DATA_IN};
endmodule
```

■ **FIGURE 5.11** Shift registers with and without a loop

Foreach loop

A foreach loop iterates over each element of an array. Foreach loops are not currently supported for synthesis, but that is likely to change in future releases of synthesizers.

In the code of Figure 5.12, a foreach loop is used to display all the values stored in a memory. Prior to that, the system task

$readmemh is used to read hexadecimal data from a text file and load it into the memory. $readmemh has no meaning for hardware design and is not synthesizable. There is also a binary version, $readmemb, which interprets data in the text file as binary rather than hexadecimal. No other radices are available in $readmem functions.

Foreach loops are part of SystemVerilog but not standard Verilog.

```
module foreachloop;
  reg [3:0] MEM [0:7];
  int i;
  initial begin
    $readmemh("memdata.txt", MEM);
    foreach(MEM[i]) $display("MEM[%d] = %h", i, MEM[i]);
  end
endmodule
```

■ **FIGURE 5.12** Foreach loop stepping through a memory array

Forever loop

Forever loops are primarily used in test fixtures to create a repeating signal such as a clock. Since forever loops do not run for a fixed number of iterations and cannot be unrolled, they are not useful for hardware design. An example of using a forever loop to create a clock signal is shown in Figure 5.13.

```
module CLKGEN(output bit CLK);
  initial begin
    CLK = 1'b0;
    forever #1 CLK = ~CLK;
  end
endmodule
```

■ **FIGURE 5.13** Clock generator using a forever loop

Anything following the forever loop in the same "initial begin" block will never run. Thus, the code of Figure 5.14 is defective. The module would compile, but CLK2 would never do anything beyond its initial assignment to zero.

```
module CLKGEN(output bit CLK, CLK2);

  initial begin
    CLK = 1'b0; CLK2 = 1'b0;
    forever #1 CLK = ~CLK;
    //Error: CLK2 will never toggle
    forever #2 CLK2 = ~CLK2;
    end
endmodule
```

■ **FIGURE 5.14**   Defective dual clock generator

Use of a fork join construct can solve the problem, as shown in Figure 5.15. The same results can be obtained by putting the forever loops in separate initial blocks.

```
module CLKGEN(output bit CLK, CLK2);

  initial begin
    CLK = 1'b0; CLK2 = 1'b0;
    fork
      forever #1 CLK = ~CLK;
      forever #2 CLK2 = ~CLK2;
    join
  end
endmodule
```

■ **FIGURE 5.15**   Two forever loops running in parallel

Repeat loop

Repeat loops always run for a fixed number of iterations. The terminating condition is fixed at the start of the loop. It is not reevaluated as the control variable changes. This is illustrated in Figure 5.16, where the loop is repeated four times, despite the control variable being updated in each iteration through the loop. The output for the module is shown below the source code. The loop will terminate after four iterations, although the loop control variable CNT will increment on every pass through the loop.

```
module repeatloop;
 initial begin
  int CNT = 4, LOOPCNT = 0;
  repeat (CNT) begin
    CNT++;
    LOOPCNT++;
    $display("CNT = %d, LOOPCNT = %d", CNT, LOOPCNT);
  end
 end
endmodule

# CNT = 5, LOOPCNT = 1
# CNT = 6, LOOPCNT = 2
# CNT = 7, LOOPCNT = 3
# CNT = 8, LOOPCNT = 4
```

■ **FIGURE 5.16** A repeat loop

Break and continue

Break and continue are methods of terminating loops early.

Continue will stop execution of the code in a loop and immediately go on to the next iteration of the loop. Break will end execution of the loop. In Figure 5.17, the memory display code of

```
module foreachloop;
  reg [3:0] MEM [0:7];
  int i;
  initial begin
   $readmemh("memdata.txt", MEM);
    foreach(MEM[i]) begin
      //skip printing on 4th & 6th words but stay in loop
      if (i == 3 || i == 5) continue;
      //quit the loop before printing the last element
      else if (i == 7) break;
      else $display("MEM[%d] = %h", i, MEM[i]);
    end
  end
endmodule
```

■ **FIGURE 5.17** Using continue and break to modify loop execution

Figure 5.12 has been modified to incorporate a break and a continue statement. When the comparison of the index to values three or five is true, the next iteration of the loop will begin without completing the display statement. When the index is seven, the loop will terminate, again without running the display statement.

Break and continue are SystemVerilog constructs and are not part of standard Verilog.

Disable

Standard Verilog has a "disable" statement that can be used for terminating loop execution. Depending on how it is called, it can act as either a SystemVerilog break or continue. It is also supposed to remove from the execution queue any assignments already placed there from the loop but not yet executed by the time the disable statement runs. Because of vagueness in the specification, disable has been implemented in different manners by different design automation companies. Disable has been superseded by break and continue. Its use is not recommended. It is not synthesizable.

Generate

Generate statements are used to create circuitry where either the quantity or type is not set until compile time. Not having the circuitry completely defined until compile time typically implies dependency on the value of a parameter.

There is some overlap between what a generate loop can do and jobs that can be accomplished with one of the other loop types. If the loop contains nothing but behavioral statements, a generate loop is not necessary, but if a variable number of instances is needed, a generate construct is the only practical method.

In Figure 5.18, a scalable binary to Gray converter is coded in two different ways. In the first module, a generate loop is used. In the second, identical circuitry is created with a for loop. Syntactical differences between the two approaches are that the control variable for a generate loop must be a "genvar" and generate statements are always external to any functional block. The begin and end statements in the generate block are needed, as the result is more than one functional statement, even though only a single assign statement has been typed in.

```
module genB2G #(SIZE = 8) (input [SIZE - 1 : 0] Bin,
  output logic [SIZE - 1 : 0] Gray);

  always_comb Gray[SIZE - 1] = Bin[SIZE - 1];
  generate
    genvar I;
    for (I = 0; I < SIZE - 1; I++)
      begin: B2G_conv
        assign Gray[I] = Bin[I] ^ Bin[I+1];
    end: B2G_conv
  endgenerate
endmodule


module loopB2G #(SIZE = 8) (input [SIZE - 1 : 0] Bin,
  output logic [SIZE - 1 : 0] Gray);

  always_comb begin
    Gray[SIZE - 1] = Bin[SIZE - 1];
    for (int I = 0; I < SIZE - 1; I++)
      Gray[I] = Bin[I] ^ Bin[I+1];
  end
endmodule
```

■ **FIGURE 5.18** A generate loop and a for loop creating identical binary to Gray converters

Current synthesis rules require that generate block loops be given a name, as is done in module genB2G, where the loop is given the name B2G_conv. Simulators are less strict and will accept unnamed loops. The target of assignment in the generate loop must not be a register type. Use of the SystemVerilog logic data type allows the same type to be used in both modules, but in standard Verilog, Gray would need to be a wire in genB2G and a reg in module loopB2G.

Another application of a generate statement is to choose between two different architectures, depending on the value of a parameter. This application is unlikely to be useful in practice, as synthesizers tend to be very good at picking an optimal algorithm based on a variety of characteristics. In the example shown in Figure 5.19, an arithmetic algorithm is chosen based solely on the size of the operands. Coding the function behaviorally and letting the synthesizer optimize for power, area, speed, and other operating environment constraints is a more practical approach than simply picking one of two based on size alone. For the code of Figure 5.19 to work, scalable models of both algorithms would also have to be coded, an additional job that is unnecessary when the function is coded behaviorally with arithmetic operators.

```
module generateselect #(WIDTH = 8)
(input [WIDTH - 1 : 0] A, B,
output logic [WIDTH : 0] SUM);

generate
  if (WIDTH < 8)
    RIPPLE  #(WIDTH)  ADDER(A, B, SUM);
  else
    CLA     #(WIDTH)  ADDER(A, B, SUM);
endgenerate
endmodule
```

■ **FIGURE 5.19** Selecting implementations based on operand size

If generate statements are not useful for either picking an algorithm or creating scalable modules, the reader may be wondering why they exist. The answer is that their utility lies in creating a scalable number of instances, as shown in Figure 5.20. Even if the number of instances is a constant, using a generate loop can turn long and unwieldy code into an efficient and easy-to-maintain module. In the example of Figure 5.20, first a scalable counter is created. Then one counter for each channel is generated. The output of the top-level module includes an array of counts. Allowing arrays as ports is a SystemVerilog enhancement. When TOPCNT is instantiated, the number of channels can be set, allowing an arbitrary number of counters to be created. Also note that while each counter has a single-bit enable signal, at the top level there is one enable bit per channel, allowing each counter to be enabled individually.

```
module COUNT #(SIZE = 4) (input CLK, RST, EN,
  output logic [SIZE - 1 : 0] CNT);

  always_ff @(posedge CLK, negedge RST)
    if (!RST) CNT <= 'b0;
    else
      if (EN) CNT <= CNT + 1;
      else CNT <= CNT;
endmodule

module TOPCNT #(CHANNELS = 2, SIZE = 4)
  (input CLK, RST, input [CHANNELS - 1 : 0] EN,
  output logic [SIZE - 1 : 0] CNT [CHANNELS - 1 : 0]);

  genvar I;
  generate
    for (I = 0; I < CHANNELS; I++) begin: CNTGEN
      COUNT #(.SIZE(SIZE))  C(CLK, RST, EN[I], CNT[I]);
    end
  endgenerate
endmodule
```

■ **FIGURE 5.20** Using a generate loop to create a variable number of instances

Multiway branching

Two types of multiway branching are included, "if" and "case." The main difference between them is that if statements always infer priority and case statements do not. Multiway branching statements are also known as decision statements or conditional statements.

If statements

If statements are so ubiquitous in Verilog that they have already been used repeatedly in the preceding examples. It is all but impossible and certainly is impractical to write high-level code without this construct. If statements are usually matched with one else and may have one or more else if clauses. Note that Verilog uses the two words fully spelled out, not "elsif," as is used in some other languages.

The conditions in if statements can be arbitrarily complex, but must eventually evaluate to a simple true or false state. The following would all be legal code:

```
if (RESET == 1'b0) //same as if (!RESET)
if (RESET) //if RESET is 1
if (X > Y)
if ((STATE == 4'b1010 && START == 1'b1) || (STATE == 4'b0101 && STOP == 1'b1))
```

The if...else if...else construct implies priority. If multiple branches are true, the first true one will be executed and the others skipped. Thus, the code for an asynchronous reset flipflop as shown in Figure 5.21 will always produce a flipflop with reset overriding clock. There is no ambiguity on priority. Priority is established by the "if" construct alone. The order of signals in the port list or sensitivity list is irrelevant.

```
module flipflop(input CLK, RST, D, output logic Q);
  always_ff @(posedge CLK, negedge RST)
    if (!RST) Q <= 0;
    else Q <= D;
endmodule
```

■ **FIGURE 5.21** Asynchronously resettable flipflop inferred with "if" statement

A one-hot to binary decoder is shown in Figure 5.22. In that code, the final "else" clause is used to catch any non-one hot values by setting the output to unknown, or Verilog X. In simulation a non-one-hot input might indicate that something has gone wrong, but to the synthesizer it means that the output is "don't care" for that clause and the most optimal circuit should be created without regard to output value.

```
module decoder1(input [7:0] DATA, output reg [2:0] CODE);
  always @(DATA)
   if (DATA == 8'b00000001) CODE = 0;
   else if (DATA == 8'b00000010) CODE = 1;
   else if (DATA == 8'b00000100) CODE = 2;
   else if (DATA == 8'b00001000) CODE = 3;
   else if (DATA == 8'b00010000) CODE = 4;
   else if (DATA == 8'b00100000) CODE = 5;
   else if (DATA == 8'b01000000) CODE = 6;
   else if (DATA == 8'b10000000) CODE = 7;
   else CODE = 3'bx;
endmodule
```

■ **FIGURE 5.22** One-hot decoder using if…else if…else statements

When synthesized with current tools, a warning is generated saying that the least significant bit of the input is unconnected. This is because the synthesizer will set the output to zero when the inputs are all zeros. With both all zeros and seven zeros followed by a single one both resulting in an output of zero, the two input patterns are indistinguishable, and the least significant bit is thus not needed.

The code of Figure 5.22 uses decimal values in the assignments to the three-bit variable CODE. This is legal in Verilog, but it is better stylistically to specify not only the value, but also the size and radix, as is done in the next example.

If the hardware design does require an indicator that the input is not one hot, some additional circuitry would need to be added. The same decoder with an error flag that is set when the input is not one hot is shown in Figure 5.23. While the CODE output by itself would not indicate an illegal value, as all possible values are legal, the ERROR flag would only be set when the input is not one hot. Because now there are multiple signal assignments in each clause, begin and end statements are also needed.

SystemVerilog has added another option for "if" clauses: set membership, which uses the new keyword "inside." Examples of

```verilog
module decoder2(input [7:0] DATA,
  output reg [2:0] CODE, output reg ERROR);
  always @(DATA)
   if (DATA == 8'b00000001) begin
     CODE = 3'd0;
     ERROR = 1'b0;
   end
   else if (DATA == 8'b00000010) begin
     CODE = 3'd1;
     ERROR = 1'b0;
   end
   else if (DATA == 8'b00000100) begin
     CODE = 3'd2;
     ERROR = 1'b0;
   end
   else if (DATA == 8'b00001000) begin
     CODE = 3'd3;
     ERROR = 1'b0;
   end
   else if (DATA == 8'b00010000) begin
     CODE = 3'd4;
     ERROR = 1'b0;
   end
   else if (DATA == 8'b00100000) begin
     CODE = 3'd5;
     ERROR = 1'b0;
   end
   else if (DATA == 8'b01000000) begin
     CODE = 3'd6;
     ERROR = 1'b0;
   end
   else if (DATA == 8'b10000000) begin
     CODE = 3'd7;
     ERROR = 1'b0;
   end
   else begin
     CODE = 3'bx;
     ERROR = 1'b1;
   end
endmodule
```

■ **FIGURE 5.23** Multiple statements in "if. . . else" clauses must be surrounded by "begin. . . end"

its use are shown in Figure 5.24. To be synthesizable, the values used in the "inside" descriptor must be constants, not variables, although wildcards are permitted. In Figure 5.24, the output HIT will be set high if DATA is 0, 2, 4, anywhere from 9 through 15 inclusive and for binary values 10000000, 10000001, 10000100, and 10000101.

```
module setinside(input [7:0] DATA, output reg HIT);
  always_comb
    if (DATA inside {0, 2, 4, [9:15], 8'b10000?0?}) HIT = 1'b1;
    else HIT = 1'b0;
endmodule
```

■ **FIGURE 5.24** Set membership conditional

For the decoders of Figures 5.22 and 5.23, the branching conditions are all mutually exclusive. The DATA input cannot simultaneously be equal to 10000000 and 01000000. With "if" statements, however, many branches may be true concurrently. This is the case in the pseudo code shown below, where nothing precludes FLOOR from being equal to 13 when an emergency occurs.

```
if (EMERGENCY) NEXT_STATE <= STOP;
else if (FLOOR == 13) NEXT_STATE <= DOWN;
else if (REQUEST13 && STATE == STOP) NEXT_STATE ...
...
else NEXT_STATE <= STATE;
```

It is because of priority that such a construct will work rather than yield nondeterministic results.

CASE STATEMENTS

The second method of multiway branching is through use of case statements. With a case statement, there is one control variable and one branch is selected according to the value of that variable. This is illustrated in Figure 5.25, where a four-to-one multiplexor is inferred with a case statement.

```
module mux4_1(A, B, C, D, SEL, OUT);
  input A, B, C, D;
  input [1:0] SEL;
  output reg OUT;
  always_comb
    case (SEL)
      2'b00: OUT = A;
      2'b01: OUT = B;
      2'b10: OUT = C;
      default: OUT = D;
    endcase
endmodule
```

■ **FIGURE 5.25** Four-to-one multiplexor using case

A decoder similar to the one using if statements of Figure 5.22 is shown in Figure 5.26.

```
module decoder3(input [7:0] DATA, output reg [2:0] CODE);
  always_comb
    case (DATA)
    8'b00000001: CODE = 0;
    8'b00000010: CODE = 1;
    8'b00000100: CODE = 2;
    8'b00001000: CODE = 3;
    8'b00010000: CODE = 4;
    8'b00100000: CODE = 5;
    8'b01000000: CODE = 6;
    8'b10000000: CODE = 7;
    default: CODE = 3'bx;
  endcase
endmodule
```

■ **FIGURE 5.26** One-hot decoder using a case statement

That case statements are limited to a single control variable and may seem to make case constructs less powerful and flexible than if statements, but the limitation can easily be finessed with the concatenation operator. By concatenating variables, any combination of control variables may be used. The four-to-one multiplexor of Figure 5.25 is reworked in Figure 5.27 to use two discrete variables concatenated together to form a two-bit selector. It works identically to the design of Figure 5.25.

```
module catcontrol(A, B, C, D, S1, S0, OUT);
  input A, B, C, D, S0, S1;
  output reg OUT;
  always_comb
    case ({S1, S0})
      2'b00: OUT = A;
      2'b01: OUT = B;
      2'b10: OUT = C;
      default: OUT = D;
    endcase
endmodule
```

■ **FIGURE 5.27** Concatenating variables to control a case statement

Two variations on the case statement allow the use of wildcards to signify don't care conditions. Casez treats bits having a Z value as don't care and casex does the same for both X and Z. The question mark can also be used to signify high-impedance state and it too will work as a don't care in casex and casez statements. In the example of Figure 5.28, when the most significant bit of the state variable is zero, the least significant bit does not matter and is not evaluated. This shorthand can produce circuit description code that is easier to understand and maintain.

```
//case 2'b0x covers both state == 00 and state == 01
always @(state)
  casex (state)
    2'b11: next_state = 2'b00;
    2'b10: next_state = 2'b11;
    2'b0x: next_state = 2'b01;
  endcase
```

■ **FIGURE 5.28** Using a wild card to set a don't care bit

A result of this dual meaning of X and Z is that they can treat unknown or high-impedance bits as don't care conditions in casex and casez statements. Errors can arise when an uninitialized signal properly takes on an X value, but the unknown signal is then treated by a casex statement as a don't care. The result can be a simulation/synthesis mismatch, as the synthesized hardware will only have zeros and ones, never any X values.

Because casez does not treat X values as don't care, some engineers prefer to use only casez, never casex. However, an unconnected/undriven signal will show up in simulation as high impedance and be treated as don't care in a casez statement, so that does not entirely eliminate the potential for such errors.

SystemVerilog has introduced a new construct, case inside, to allow don't care inputs to case statements without the unintended consequences of casex and casez. With case inside, unknown or high-impedance inputs are not treated as don't care, as they are in casex and casez, allowing error conditions to be more easily detected and debugged. The difference between case inside and casex is illustrated in Figure 5.29. In that example, the control input is not initialized at time zero, so it starts off unknown. The casex construct in the second module treats unknown bits as don't care conditions, so the first branch is selected. With the case inside construct of the first module, 3'bXXX does not match any of the enumerated clauses, so the default branch is taken.

While the distinction is subtle, current best practice recommendations are to use case inside rather than casex or casez.

Begin and end statements are not needed in any of the preceding case examples, but they are needed whenever more than one action is to be taken in any case clause. Figure 5.30 has a snippet of code showing this.

A defective attempt to achieve the same results is shown in Figure 5.31. In that code, only one assignment in each case will be performed. The other will be skipped.

```
module insidedemo(input [2:0] OPCODE,
  output logic [3:0] ENABLE_BUS);
  always_comb
    case (OPCODE) inside
      3'b0xx: ENABLE_BUS = 4'd0;
      3'b100: ENABLE_BUS = 4'd1;
      3'b101: ENABLE_BUS = 4'd2;
      default:  ENABLE_BUS = 4'd7;
    endcase
endmodule

module casexdemo(input [2:0] OPCODE,
  output logic [3:0] ENABLE_BUS);
  always_comb
    casex (OPCODE)
      3'b0xx: ENABLE_BUS = 4'd0;
      3'b100: ENABLE_BUS = 4'd1;
      3'b101: ENABLE_BUS = 4'd2;
      default:  ENABLE_BUS = 4'd7;
    endcase
endmodule

module tb_insidedemo;
  reg [2:0] OPCODE;
  logic [3:0] ENABLE_BUS1, ENABLE_BUS2;
  insidedemo  UUT1(OPCODE, ENABLE_BUS1);
  casexdemo   UUT2(OPCODE, ENABLE_BUS2);
  initial begin
    $monitor("OPCODE = %b, BUS1 = %d, BUS2 = %d", OPCODE, ENABLE_BUS1, ENABLE_BUS2);
    OPCODE = 3'bx;
    #1 OPCODE = 3'b0xx;
    #1 OPCODE = 3'b000;
    #1 OPCODE = 3'b001;
    #1 OPCODE = 3'b100;
  end
endmodule

/*Results: difference at time 0. Unknown opcode
takes the default clause with case inside but
matches the first branch for casex.
# OPCODE = xxx, BUS1 =  7, BUS2 =  0
# OPCODE = 0xx, BUS1 =  0, BUS2 =  0
# OPCODE = 000, BUS1 =  0, BUS2 =  0
# OPCODE = 001, BUS1 =  0, BUS2 =  0
# OPCODE = 100, BUS1 =  1, BUS2 =  1
```

■ **FIGURE 5.29** Using case inside to avoid treating don't know bits as don't care

```
always_comb
  case (ALPHA)
  0: begin
    OUT1 = A;
    OUT2 = B;
  end
  1: begin
    OUT1 = C;
    OUT2 = D;
  end
```

■ **FIGURE 5.30** Multiple assignments need to be surrounded by begin...end

```
//This one DOES NOT WORK!!!

always_comb
  case (ALPHA)
    0: OUT1 = A;
    0: OUT2 = B;
    1: OUT1 = C;
    1: OUT2 = D;

//Only 1 of each will happen.
```

■ **FIGURE 5.31** Erroneous attempt to execute multiple assignments per case

With if statements, any number of branches may be simultaneously true without ambiguity. Case statements can also be coded so that more than one branch is true for a given control value. This is known as overlapping cases. In the absence of any other directives, overlapping cases are interpreted as inferred priority, with the clauses given priority from top to bottom. In the example of Figure 5.32, when CONTROL is 1111, the ALPHA will be set to one, and BETA will be set to zero.

```
module overlap(CONTROL, ALPHA, BETA);
   input [3:0] CONTROL;
   output reg ALPHA, BETA;
   always_comb
     /*both enumerated clauses will be true
     for CONTROL = 1111*/
     casez (CONTROL)
       4'b11??: begin
         ALPHA = 1;
         BETA = 0;
       end
       4'b??11: begin
         BETA = 1;
         ALPHA = 0;
       end
       default: begin
         ALPHA = 0;
         BETA = 0;
       end
     endcase
endmodule
```

■ **FIGURE 5.32** Overlapping case branches

When synthesized, priority circuitry will be generated. Priority circuitry can be far larger and more complex than a fully parallel design, when no two branches may be simultaneously true.

LATCH GENERATION

So far all conditional statements have been completely specified. When they are not, latches are inferred. This is true for both "if" and "case" structures.

If conditionals are completely specified when each has an else clause. Case conditionals are completely specified when each has a default clause. Because of X and Z values, case conditionals may not be considered to be completely specified by all tools even if all possible binary values are enumerated in the absence of a default clause.

In the example of Figure 5.33, an incompletely specified "if" construct is shown. It explicitly says what to do if CONTROL is logic one. It does not say what to do otherwise. If synthesized, this code would result in a D latch being generated.

```
/*Latch D when CONTROL is high. Note that, unlike
a D flipflop, D must be in the sensitivity list
when using classic Verilog design style*/

module DLATCH(input D, CONTROL, output reg Q);
  always @(CONTROL or D)
    if (CONTROL) Q = D;
endmodule
```

■ **FIGURE 5.33** D latch with an incompletely specified if clause

Taking advantage of SystemVerilog extensions, the same D latch code can be written with an always_latch construct, eliminating the need to have a sensitivity list and instructing the compiler to issue a warning if the code does not infer a latch. This is shown in Figure 5.34.

```
/*Latch D when CONTROL is high, SystemVerilog style*/

module DLATCH(input D, CONTROL, output reg Q);
  always_latch
    if (CONTROL) Q = D;
endmodule
```

■ **FIGURE 5.34** D latch using an always_latch construct

A similar result can be obtained with a case statement. In the code of Figure 5.35, a two-bit latched bus will be created.

Latches are rarely desired in modern digital designs. Usually a netlist incorporating a latch is an error, not a design technique. Use of a default clause with every case statement and an else clause with every if construct will always avoid latch generation.

```
module latchstate(STATE, NEXT_STATE);
  input [1:0] STATE;
  output reg [1:0] NEXT_STATE;

  localparam FETCH = 2'b00, DECODE = 2'b01, EXECUTE = 2'b10;

  /*This will create a two-bit latch. When STATE is 11,
  the previous value of NEXT_STATE will be held.*/
  always @(STATE)
  case (STATE)
    FETCH: NEXT_STATE = DECODE;
    DECODE: NEXT_STATE = EXECUTE;
    EXECUTE: NEXT_STATE = FETCH;
  endcase
endmodule
```

■ **FIGURE 5.35** D latches with an incompletely specified case statement

Another technique for avoiding latch generation is shown in Figures 5.36 and 5.37. In these examples, an unconditional assignment is made before reaching the multiway branching structure. If none of the branches is taken, there is still an assignment to the variable, so it will not be latched. This technique is not accepted by all tools. Use of else and default clauses to avoid latches are the recommended methodology.

```
module nolatches(input [2:0] OPERATOR,
  input [7:0] ICOUNT, PCOUNT, XCOUNT,
  output reg [7:0] DATA);
  always_comb begin
    DATA = 8'b0; //default assignment
    case (OPERATOR)
      3'b001: DATA = ICOUNT;
      3'b010: DATA = PCOUNT;
      3'b100: DATA = XCOUNT;
    endcase
  end
endmodule
```

■ **FIGURE 5.36** Avoiding latches with an unconditional assignment prior to multiway branch

```
module nolatches2(input [2:0] OPERATOR,
  input [7:0] ICOUNT, PCOUNT, XCOUNT,
  output reg [7:0] DATA);
  always_comb begin
    DATA = 8'b0; //default assignment
    if (OPERATOR == 3'b001) DATA = ICOUNT;
    else if (OPERATOR == 3'b010) DATA = PCOUNT;
    else if (OPERATOR == 3'b100) DATA = XCOUNT;
  end
endmodule
```

■ **FIGURE 5.37** Alternative method of avoiding latch generation

UNIQUE AND PRIORITY

Synthesis tools have long incorporated pragmas, or synthetic comments, that instruct the synthesizer as to the intent of the designer. These are statements that look like comments to simulators but are read by synthesizers to influence the circuit design at the gate level. Using them is extremely problematic, as they can cause synthesis/simulation mismatches [1].

To allow designers to continue to instruct tools as to the desired functionality but without simulation/synthesis mismatch risk, SystemVerilog has introduced two modifiers to case and if statements that are interpreted equally by simulators and synthesizers.

When the keyword "unique" is added to a multiway branching statement, the tools are informed that no two branches are supposed to be simultaneously true, that is, there are no overlapping cases. If simulation finds that there are overlapping cases, a warning or error will be reported. In synthesis, use of "unique" instructs the tools not to make a priority encoder. This is true for both types of branch constructs. Because synthesizer algorithms can detect fully parallel conditions even in the absence of the "unique" keyword and optimize their output accordingly, there is often no difference between the synthesis results of code with and without the unique modifier. This is the case for the code of Figure 5.38, which produced identical circuitry when "unique" was removed. While use of the "if" structure indicates that a priority encoder is wanted, in this case a smaller, faster, fully parallel decode will work identically and is thus generated.

```
module uniquedemo (input [7:0] DATA, output logic [2:0] CODE);
  always_comb
   unique if (DATA == 8'b00000001) CODE = 0;
   else if (DATA == 8'b00000010) CODE = 1;
   else if (DATA == 8'b00000100) CODE = 2;
   else if (DATA == 8'b00001000) CODE = 3;
   else if (DATA == 8'b00010000) CODE = 4;
   else if (DATA == 8'b00100000) CODE = 5;
   else if (DATA == 8'b01000000) CODE = 6;
   else CODE = 7;
endmodule
```

■ **FIGURE 5.38** "Unique" instructs tools that there should be no overlapping conditions

In the code of Figure 5.39, "unique" was added to the overlapping case conditions of Figure 5.32. Since the clauses are not mutually exclusive, compilation of the code will produce a warning in simulation. In synthesis, the parallel decoder produced with the

```
module overlap2(CONTROL, ALPHA, BETA);
  input [3:0] CONTROL;
  output reg ALPHA, BETA;
  always_comb
    /*Clauses are overlapping but adding unique
    modifier instructs synthesizer not to make a
    priority encoder.*/
    unique casez (CONTROL)
      4'b11??: begin
        ALPHA = 1;
        BETA = 0;
      end
      4'b??11: begin
        BETA = 1;
        ALPHA = 0;
      end
      default: begin
        ALPHA = 0;
        BETA = 0;
      end
    endcase
endmodule
```

■ **FIGURE 5.39** Using "unique" modifier to force parallel circuit synthesis

unique version is smaller and faster than the original (nonunique) encoding. This is shown in Figures 5.40 and 5.41.

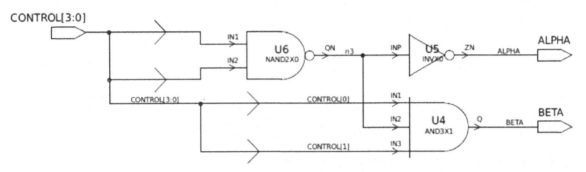

■ **FIGURE 5.40** Synthesized priority circuit (no "unique")

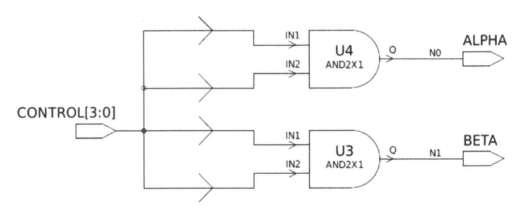

■ **FIGURE 5.41** Synthesized parallel circuit ("unique casez" in source code)

The other SystemVerilog modifier is "priority." When added to a case statement, it signals that all possible legal values for the control variable have been enumerated. In simulation, if a control signal value is encountered that does not match any of the case clauses, an error will be reported. In synthesis, the result of adding a priority modifier is that latches will not be generated for any missing control values. If a case statement has a default clause, all possible values are covered, so addition of a priority modifier will be without effect, as shown in Figure 5.42. The designer knows that both EN1 and EN2 can never be active simultaneously, so that condition is left out. The synthesized circuit will not have any latches. If both inputs are simultaneously asserted during verification, a warning or error report will be generated.

```
module priority_demo(input EN1, EN2,
  output logic ALPHA, BETA);
  always_comb
    priority case ({EN2, EN1})
      2'b00: begin
        ALPHA = 0;
        BETA = 0;
      end
      2'b01: begin
        ALPHA = 1;
        BETA = 0;
      end
      2'b10: begin
        ALPHA = 0;
        BETA = 1;
      end
    endcase
endmodule
```

■ **FIGURE 5.42** Latch avoidance with "priority" modifier

Using priority with a case statement does not ensure that no latches will be generated under all conditions. In the code of Figure 5.43, output BETA will still be latched because no assignment is made to it when the control input is 2'b10. Even adding a default clause will not prevent the latch from being included in the synthesized circuit.

```
module priority_demo(input EN1, EN2,
  output logic ALPHA, BETA);
  always_comb
    priority case ({EN2, EN1})
      2'b00: begin
        ALPHA = 0;
        BETA = 0;
      end
      /*What about BETA?
      A latch will be generated!
      The priority modifier prevents
      ALPHA from being latched but the
      missing value from this enumerated
      clause will cause a latch to be
      generated for BETA.*/
      2'b01: begin
        ALPHA = 1;
      end
      2'b10: begin
        ALPHA = 0;
        BETA = 1;
      end
      2'b11: begin
        ALPHA = 1;
        BETA = 1;
      end
    endcase
endmodule
```

■ **FIGURE 5.43** Priority modifier does not prevent all latches

■ **SUMMARY**

Loops and branches are the essence of high-level design. Their judicious use is essential to take advantage of the power offered by circuit description using hardware description languages. Some care must be taken to avoid generating unwanted latches when using branching statements.

Most looping constructs are synthesizable, although loops are more commonly found in test fixtures than in circuit description.

When loops are used in circuit descriptions, their end points must be fixed at compile time. The number of iterations through a loop must not be data dependent in a circuit description, although there is no such restriction in test fixtures.

REFERENCE

[1] Cummings, CE "SystemVerilog's Priority & Unique – A Solution to Verilog's 'ull_case' & 'parallel_case' Evil Twins!" SNUG Israel, 2005.

Subroutines and interfaces

Verilog has two types of subroutines, tasks and functions. Each has its limitations and advantages, which will be discussed in this chapter. The most fundamental difference between tasks and functions is that functions cannot contain any type of delay, whereas tasks do not share this restriction.

SystemVerilog also has introduced interfaces, which offer another place where common code can be offloaded to reduce module complexity and simplify maintenance.

Instances of modules are not considered to be subroutines, even though they act a lot like them. A common design technique is to make one scalable model of commonly used components such as counters, then instantiate as needed. This sounds a lot like calling a subroutine. Nevertheless, in the formality of the language, an instance is not a subroutine and a subroutine is not an instance.

Whether tasks, functions, or instances, the use of any of these constructs can simplify a design by breaking complex procedures into smaller, more manageable, and reusable blocks. Keeping modules short and focused is always good technique. Judicious use of subroutines can prevent a module from becoming a rambling, incomprehensible mess.

SUBROUTINES

Both task and function definitions must be made within a module. Including all the subroutine code in the module can cause the module to become as long and unwieldy as if subroutines were not used in the first place. However, the actual subroutine code can be kept in a separate file and the file referenced in an imported package or with an "include directive," keeping each file concise and focused.

Subroutines define a new level of hierarchy in the design, but unlike modules, tasks and functions do not have instance names and are referenced by their design name. For example, if task T_TIME has a local variable called X23 and the task is enabled in instances A and B of design MID, which is instantiated in design TOP, the X23 variables would be referenced from TOP as A.T_TIME.X23 and B.T_TIME.X23. Module instances need instance names, but subroutine calls do not.

One key distinction between subroutines and instances is that subroutines cannot contain any instantiations, whereas instantiated modules may contain other instantiations, enable tasks, and call functions.

TASKS

Tasks are enabled when the name of the task is encountered in code. They can be enabled from another task or a module but not from a function. This restriction is needed because tasks can contain delays but functions cannot.

Tasks can have input, output, and bidirectional ports, just like modules. They can have local variables. They can also operate on variables of the enabling module or task. When using ports, with standard Verilog calls to the task must provide arguments in the same order in which they are declared in the task. SystemVerilog allows named port association.

In early versions of Verilog, variable allocation for tasks and functions was static. In simulation, space for each variable of the subroutine was allocated when compiled. This meant that multiple, concurrent calls to a task or function would lead to data corruption, as the same physical memory locations would be referenced for the overlapping subroutine reads and writes. To solve this problem, "automatic" tasks and functions were added to the language. Automatic subroutines use dynamic memory allocation, allowing concurrent calls to subroutines to reference discrete memory locations rather than overwriting the locations used by other calls to the same subroutine.

To be synthesizable, subroutines defined in packages must always be automatic. Subroutines defined directly in a module do not share this restriction.

An example of using a task as a clock generator is shown in Figure 6.1. Using a repeat loop, the task will toggle the clock as many times as indicated by the input variable TIMES. This task is written in standard Verilog. SystemVerilog allows port declarations following the task name, as is done with module declarations. An example of that style is shown in Figure 6.5.

```
task TOGGLE;
  input integer TIMES;
  repeat (TIMES) begin
    #(PERIOD / 2) CLK = 1'b0;
    #(PERIOD / 2) CLK = 1'b1;
  end
endtask
```

■ **FIGURE 6.1** A task used to create a clock signal

Task TOGGLE references the variable CLK, which is not local to the task. Operation of the task depends on the enabling module having a CLK variable and having defined parameter PERIOD.

Tasks can be enabled in parallel with all other events in a module or in series with other Verilog assignments and other subroutines. This is illustrated in Figure 6.2, where there are two calls to task TOGGLE, one of them commented out.

```
module taskdemo;
  parameter WIDTH = 4;
  parameter PERIOD = 10.0;
  reg [WIDTH - 1 : 0] COUNT, DATA;
  reg CLK, RST, LOAD;

  always @(posedge CLK or negedge RST) begin
    if (!RST) COUNT <= 'b0;
    else begin
      if (LOAD) COUNT <= DATA;
      else COUNT <= COUNT + 1;
    end
  end

  initial TOGGLE(12); //Parallel execution

  initial begin
    //TOGGLE(12); //Serial execution
    DATA = 4'ha;
    RST = 1'b1; LOAD = 1'b0;
    #PERIOD RST = 1'b0;
    #(2*PERIOD) RST = 1'b1;
    #(2*PERIOD) LOAD = 1'b1;
    #PERIOD LOAD = 1'b0;
  end

  task TOGGLE;
    input integer TIMES;
    repeat (TIMES) begin
      #(PERIOD / 2) CLK = 1'b0;
      #(PERIOD / 2) CLK = 1'b1;
    end
  endtask
endmodule
```

■ **FIGURE 6.2** Calling task TOGGLE in series and in parallel

Simulation of parallel execution of the task is shown in Figure 6.3. It works as intended, generating 12 clock cycles, which are used to increment, load, and increment again a counter.

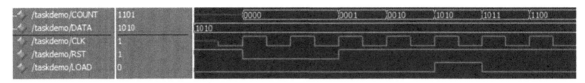

■ **FIGURE 6.3** Simulation of parallel call to task TOGGLE

In Figure 6.4, a simulation of serial execution of the task is shown. Again, the task causes a clock signal to be generated, but this time the counter does not count. In this setup, task TOGGLE is enabled, but it blocks execution of all other signal assignments in its block until it returns, which is after it has generated all 12 clock pulses. Only after the clock stops do other signal assignments, including reset, take place.

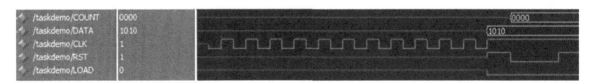

■ **FIGURE 6.4** Simulation of a serial call to task TOGGLE

In Figure 6.5, task TOGGLE is modified to have a local variable CLK, which is then set to be the output of the task.

```
task TOGGLE;
  input integer TIMES;
  output reg CLK;
  repeat (TIMES) begin
    #(PERIOD / 2) CLK = 1'b0;
    #(PERIOD / 2) CLK = 1'b1;
  end
endtask
```

■ **FIGURE 6.5** Task with a local variable

Figure 6.6 shows module taskdemo modified to use the new task. Instead of just passing it the number of times it should cycle the clock, it also has an output port for the local clock. When there

```
module taskdemo;
  parameter WIDTH = 4;
  parameter PERIOD = 10.0;
  reg [WIDTH - 1 : 0] COUNT, DATA;
  reg CLK, RST, LOAD;

  always @(posedge CLK or negedge RST) begin
    if (!RST) COUNT <= 'b0;
    else begin
      if (LOAD) COUNT <= DATA;
      else COUNT <= COUNT + 1;
    end
  end

//Task now has both an input and an output
  initial TOGGLE(12, CLK); //Parallel execution

  initial begin
    //TOGGLE(12); //Serial execution
    DATA = 4'ha;
    RST = 1'b1; LOAD = 1'b0;
    #PERIOD RST = 1'b0;
    #(2*PERIOD) RST = 1'b1;
    #(2*PERIOD) LOAD = 1'b1;
    #PERIOD LOAD = 1'b0;
  end

  task TOGGLE; //Operates on local variable CLK
    input integer TIMES;
    output reg CLK;
    repeat (TIMES) begin
      #(PERIOD / 2) CLK = 1'b0;
      #(PERIOD / 2) CLK = 1'b1;
    end
  endtask

endmodule
```

■ **FIGURE 6.6** Call to task TOGGLE with a local variable

was no local CLK variable, the task operated on the module CLK. Now that there is a local one, that one will be used.

Simulation of the new system is shown in Figure 6.7. It does not work as intended. The problem is that the local task variable CLK is independent of the module variable CLK. The two are not associated. The task clock toggles as instructed, but the clock that is connected to the counter is never driven at all and stays unknown. Because the task is called in parallel to the other signal assignments, the reset signal does work as intended, setting the counter to all zeroes. It never moves beyond that, as it is never clocked.

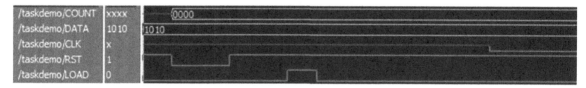

■ **FIGURE 6.7** Task local variable CLK is not associated with module variable CLK

Simulation of the new task in series yields the waveform of Figure 6.8. Once again, the task clock toggles 12 times, but this time it prevents the other signals from being assigned until the task is through. Only then does the reset take place.

■ **FIGURE 6.8** Simulation of serial call to task TOGGLE with local variable CLK

The code with the local variable CLK can be made to work by replacing the reference to the module clock with a hierarchical reference to the task clock, as shown in the line of code below.

```
always_ff @(posedge TOGGLE.CLK, negedge RST)
```

While internally consistent, it offers no improvement over the first version, where the task does not have a local clock variable.

Task TOGGLE could be used as part of a test procedure, but it would not produce any hardware if synthesized. The delays would be ignored in synthesis, rendering the whole task meaningless from a hardware perspective.

Tasks can contain both explicit and implicit delays, but they must have some specified means of terminating. The implication of this is that no task may contain an always or initial block. In the example of Figure 6.9, there are explicit delays (wait for one and three time units) and an implicit delay (wait for the rising edge of CLK). Changing

> @(posedge CLK)

to

> always @(posedge CLK)

would be illegal and the task would not compile, as adding "always" turns the implicit delay into a never-ending functional block, and tasks must have a way of ending, even if the end can be delayed forever by waiting for an event that never happens.

@(UNOBTANIUM) //unobtainium may never happen, but if it does, the task will end

```
logic POSITIVE, NEGATIVE;
logic [15:0] DATABUS;

task automatic PARITY([15:0] DBUS, output POSP, NEGP);
  @(posedge CLK)
    #1 POSP = ^DBUS; //output positive parity
    #3 NEGP = ~^DBUS; //output negative parity
endtask

always @(DATABUS)
  PARITY(DATABUS, POSITIVE, NEGATIVE);
```

■ **FIGURE 6.9** Delays in a task

Simulation of the code shown in Figure 6.9 illustrates a difference between enabling a task and instantiating a module. Both parity bits in the enabling module will be set four time units after the rising edge of the clock. This is because the task does not return values as they are set in the task. Values are returned after the task ends, which takes $(1 + 3)$ time units.

Synthesizers ignore explicit delays, so whatever values are included in whatever order have no impact on the resulting circuit. The task of Figure 6.9 would not be synthesizable with current tools, which do not allow event control statements in tasks, but it can be compiled and simulated.

A sequential circuit using a task can be synthesized by moving the edge control statement to the module, as shown in Figure 6.10. The task itself only represents combinational circuitry.

```
logic POSITIVE, NEGATIVE;
logic [15:0] DATABUS;

task automatic PARITY([15:0] DBUS, output POSP, NEGP);
    #1 POSP = ^DBUS; //output positive parity
    #3 NEGP = ~^DBUS; //output negative parity
endtask

always_ff @(posedge CLK)
  PARITY(DATABUS, POSITIVE, NEGATIVE);
```

■ **FIGURE 6.10** Synthesizable sequential circuit with a task

Tasks may be synthesizable but they find more use in test fixtures than in circuit descriptions. A typical usage would be to apply a sequence of values that includes control and data signals to a device under test.

An example of offloading a repetitive sequence of assignments to a task is shown in Figure 6.11. In that example, a handshaking protocol is implemented in a task. Using the task, long sequences of tests can be implemented by calling the task with new values for the address and data buses, eliminating the need to include code for the handshake protocol with every new test value. The task uses wait statements, which are not synthesizable, to suspend operation until the indicated signals are set to logic one. The same effect could be accomplished with more @ statements. In this example, internally generated signals are set synchronously with the clock, but response to the external signals READY and ACK is asynchronous. Since this is part of a test program and not circuit design, the normal protocols for synchronous design are irrelevant.

```
task handshake();
  wait (READY);
  @(posedge CLK)
    VALID = 1'b1;
    DATA = RAWDATA;
    ADDRESS = RAWADDR;
  wait (ACK);
  @(posedge CLK) VALID = 1'b0;
endtask
```

■ **FIGURE 6.11** Using a task to execute repetitive series of commands

With SystemVerilog, the need to enclose a series of commands with "begin…end" was dropped for subroutines. The beginning and end are implicit, and evaluation of the statements will be sequential. Nonsynthesizable tasks can use "fork…join" to force parallel evaluation. SystemVerilog also has task ports default to input in the absence of an explicit declaration and has ports take the same type as the last declaration of direction and size.

In the SystemVerilog example of Figure 6.10, DBUS will default to input and both POSP and NEGP will be single-bit outputs. If the two signal assignments were in a module, they would need to be surrounded by a begin and end statement, but those can be left out in a SystemVerilog task. The task is automatic, which allows it to be imported from a package. Since the task does not contain any delays, static memory allocation will work as well as dynamic for simulation, so the original reason for making tasks automatic does not come into play in this case.

SystemVerilog has added "return" statements to subroutines. When a return statement is encountered, the routine will end. This is another new feature unlikely to be useful in circuit design but commonly found in verification suites. A task using returns in an error checker is shown in Figure 6.12. This sort of technique is often used to reduce the time spent on useless simulation after an error has been detected. In the task shown, both buses are inputs and FLAG is a module variable, so it does not need to be declared in the task. The task has no outputs other than display statements.

```
task look_for_X([31:0] DATA_BUS, [23:0] ADDR_BUS);
  if (DATA_BUS === 'bZ) begin
    $display("Data Bus is high impedance");
    return;
  end
  else if (DATA_BUS === 'bX) begin
    $display("Data Bus is unknown");
    return;
  end
  else begin
    case (ADDR_BUS)
      0: if (DATA_BUS != 32'h0) FLAG = 1'b1;
      //rest of test cases go here
    endcase
  end
endtask
```

■ **FIGURE 6.12** Task with return statements

FUNCTIONS

The biggest single difference between tasks and functions is that functions cannot contain delays of any type. They all must run in zero simulation time. This restriction means that in circuit design, functions can only be used to infer combinational circuitry, although of course circuitry synthesized from a function will incur normal gate delays.

A consequence of this rule is that functions can only use the blocking assignment operator. This restriction is sensible, as the reason for the existence of the nonblocking assignment operator is to ensure correct operation of sequential circuit descriptions.

Because tasks do not share this restriction, functions cannot enable tasks, although tasks can call functions. Both tasks and functions can be called from modules, although neither tasks nor functions can instantiate modules or primitive operators.

Functions in standard Verilog were required to return exactly one value. SystemVerilog adds the option to have multiple outputs from a function. The single-value restriction was never a serious limitation, since the concatenation operator could be used to string together as many operands as needed for the return value of a function.

SystemVerilog also adds void functions, which do not return any value. Examples of void functions for use in both verification and circuit design are included in this section.

The positive- and negative-parity generator of Figure 6.10 is implemented as a function in the code of Figure 6.13. In this code, the two parity variables are concatenated together to form the return value of the function, which is then assigned to two concatenated variables in the calling module. The size of a function is the number of return bits, which in this case is two. In the absence of any range, a function will return a single bit. The range may be

```
module pargen (input [15:0] DATA, output logic POSP, NEGP);

  function automatic [1:0] calcparity;
    input [15:0] BUS;
    logic P, N;
    P = ^BUS;
    N = ~^BUS;
    calcparity = {P, N};
  endfunction

  always_comb {POSP, NEGP} = calcparity(DATA);
endmodule
```

■ **FIGURE 6.13** A function returning two one-bit values concatenated together

implicitly set by specifying a multibit variable type, such as byte or integer. An example of this is shown in Figure 6.14, where the function returns a 1-byte value, which is a reordered version of the data input byte. It is the size declaration alone of the function that determines how many bits are returned.

```
function byte scramble;
  input byte DIN;
  for (int I = 0; I < 8; I++)
    if (I % 2)
      scramble[I] = DIN[I/2];
    else
      scramble[I] = DIN[I/2 + 4];
endfunction
```

■ **FIGURE 6.14** A function returning an eight-bit value

With SystemVerilog, functions also gained the capacity to have outputs and bidirectional ports. The parity generator function is reworked in Figure 6.15 to use outputs rather than to assign the generated values to the function. Because the function does not return a value, it is declared to be type void. Void functions are commonly used in verification to initiate processes, but function calcp2 is synthesizable and could be used in a circuit design. As is the case with instantiated modules, the formal arguments in the call to the function may be different from the names used in the function itself. In Figure 6.15, module signals DATA, POS, and NEG are mapped to function ports BUS, P, and N, respectively.

```
function automatic void calcp2(input [15:0] BUS, output logic P, N);
  P = ^BUS;
  N = ~^BUS;
endfunction

logic POS, NEG;
logic [15:0] DATA;
always_comb calcp2(DATA, POS, NEG);
```

■ **FIGURE 6.15** A function with outputs and a call to that function

A standard Verilog function must have at least one formal argument, but SystemVerilog has eliminated that requirement. A function that has no arguments, no inputs, no outputs, and returns no values could not only be legal, but often finds a place in verification suites. An example of such a function is shown in Figure 6.16, where void function JUSTDISPLAY is called by module JDF to display an error message when two values in the calling module do not match.

```
function void JUSTDISPLAY;
  $display("ERROR: Unexpected Output Value");
endfunction

module JDF;
  reg A, B;
  initial begin
    A = 0; B = 0;
    #1 B = 1;
    #1 A = 1; B = 0;
    #1 B = 1;
    #1;
  end

  always @(A, B)
    if (A != B) JUSTDISPLAY;
endmodule
```

■ **FIGURE 6.16** A void function having no arguments or return value

A slight modification of the void function is presented in Figure 6.17. In that code, the function does have two formal arguments, integer inputs A and B. In function ERRORCHECK, use of formal arguments allows the function to present a more meaningful message and offloads the logic of checking the variables to the function.

```
function void ERRORCHECK(input integer A, B);
  if (A != B) $display("ERROR: at time %d A = %b, B = %b", $time, A, B);
endfunction
```

■ **FIGURE 6.17** A void function with arguments but no outputs or return value

Because a task can call a function but a function cannot enable a task, putting reusable code that does not return any value into a void function is generally preferable to putting it in a task. This allows future users to call the function from another function if that proves to be convenient, whereas putting the code in a task would make that impossible.

Like tasks, SystemVerilog functions too can have return statements. Their use can speed up simulation but is unlikely to provide any benefit for synthesized hardware. An example of using return statements to terminate a function early is shown in Figure 6.18. The function in this example returns the factorial of an input value. To accomplish this, the function recursively calls itself. Recursion is an inherently nonsynthesizable construct, as the number of iterations is data dependent. This function will not work for all values even in simulation, as the longest Verilog variable type, the 64-bit longint, is nowhere near long enough to hold the factorial of the largest possible input value.

```
module returndemo(input byte DATA_IN, output longint ANSWER);

  function automatic longint factorial(input byte DATA_IN);
    if (DATA_IN < 2) return 1;
    else if (DATA_IN == 2) return 2;
    else factorial = DATA_IN * (factorial(DATA_IN - 1));
    endfunction

  always_comb ANSWER = factorial(DATA_IN);
endmodule
```

■ **FIGURE 6.18** A function with return statements

Functions have an advantage over tasks in that they can be called within other Verilog expressions. In the following example, a parity function is called and the result used all in a single line. Because the parity function is so simple, there is not really any benefit in using a function, but the same technique can be applied to functions of any level of complexity. Use of the nonblocking assignment operator implies that this code is to be included in a clocked block. While functions cannot use the nonblocking assignment operator, nothing precludes functions from being called from a line in which that operator is used.

```
if (parity(DBUS[15:0]) != DBUS[16]) PFLG <= 1'b1;
else PFLG <= 1'b0;
```

PARAMETERS IN SUBROUTINES

As is the case with packages, parameters declared in subroutines act like localparams. There is no mechanism for overriding their initial values. This does not mean, however, that subroutines are forever stuck with fixed values. Just as an undeclared variable in a subroutine will operate on a module variable, module parameters can be referenced in tasks and functions. The solution to making scalable subroutines is to simply depend on parameters in the calling or enabling module.

In Figure 6.19, a function is called with code that looks like a parameter override. However, it is actually a delay. If the default value of 32 is overridden for module paramfunc, the new value will be used in calc_parity, too. When simulated, parity will be

```
module paramfunc(DATA, PARITY);
  parameter WIDTH = 32;
  input [WIDTH - 1 : 0] DATA;
  output reg PARITY;

  //#8 is a delay, not a parameter redefinition
  always @(DATA) PARITY = #(8) calc_parity(DATA);

function calc_parity; //Uses module parameter
  input reg[WIDTH - 1 : 0] DATA; //Width is 32 here too
  calc_parity = ^DATA;
endfunction

endmodule
```

■ **FIGURE 6.19** Function using a parameter from the calling module

returned only after a delay of eight time units, but it will be calculated over the entire 32-bit data bus.

In the defective code of Figure 6.20, the function parameter will not be overridden. Instead of calculating parity over the entire data bus, it will be calculated only over the eight least significant bits and that only after a delay of 32 time units. There will be no indication that the returned parity bit covers only one quarter of the intended target.

```
module paramfunc2(DATA, PARITY);
  parameter WIDTH = 32;
  input [WIDTH - 1 : 0] DATA;
  output reg PARITY;

always @(DATA) PARITY = #(32) calc_parity(DATA);

//Local declaration of WIDTH will not be over-ridden
function calc_parity;
  parameter WIDTH = 8;
  input reg[WIDTH - 1 : 0] DATA; //Now WIDTH is 8.
  calc_parity = ^DATA; //Parity calculated only over 8 bits
endfunction

endmodule
```

■ **FIGURE 6.20** Defective use of a parameter in a function

An alternative to using the module parameters is to create a scalable virtual class in SystemVerilog, then defining the class in the $unit declaration space, a fraught exercise that will not be covered further.

MANAGING SUBROUTINES

To keep files down to a manageable size, subroutines may be put into separate files, but then they must be included or imported. Importing subroutines from packages may seem like the most elegant solution, but it is not always practical. This is because a package containing subroutines that have dependencies on module parameters will not compile. An example of this is shown in Figures 6.21 and 6.22. The package (Figure 6.21) will not compile because the referenced parameters and variables are undefined before the module is compiled. The module (Figure 6.22) will not compile because it can't find the package, which fails to compile because the variables and parameters from the module are undefined. Stalemate.

```
package parpackage;
  //depends on module definitions: will not compile
  function automatic void parity(input [SIZE - 1 : 0] DATA,
    output logic P, N);
      P = ^DATA;
      N = ~^DATA;
  endfunction
endpackage
```

■ **FIGURE 6.21** Package with a parameter. Will not compile

```
/*This does not work as written. It can easily be modified to
use include, which will then work.*/
module include_or_import #(SIZE = 8) (input [SIZE - 1 : 0] DATA,
  output logic POS, NEG);
  import parpackage::*;
  always_comb parity(DATA, POS, NEG);
endmodule
```

■ **FIGURE 6.22** Module will fail to compile when it fails to find the package

There is a partial workaround using SystemVerilog references, but, as will be shown shortly, that too has a limitation.

Fortunately, there is also an easy workaround. The subroutines can still be offloaded into a separate file and then included. In standard Verilog, the syntax requires using the "include directive" and enclosing the filename (and path, if the file is not in the current working directory) in quotation marks, but the example shown in Figure 6.23 works as written in SystemVerilog. This

```
module include_or_import #(SIZE = 8) (input [SIZE - 1 : 0] DATA,
  output logic POS, NEG);
  //include, don't import, when subroutine has dependencies
  include parpackage.sv;
  always_comb parity(.D(DATA), .P(POS), .N(NEG));
endmodule

//contents of file parpackage.sv: keywords package & endpackage removed,
//making it no longer a package
  function automatic void parity(input [SIZE - 1 : 0] D, output logic P, N);
    P = ^D;
    N = ~^D;
  endfunction
```

■ **FIGURE 6.23** Included a subroutine from a different file

example also uses named port association for the function, which is legal in SystemVerilog but not in standard Verilog.

SystemVerilog has added references, which allow subroutines in packages to use module variables. References are used in the code of Figure 6.24 to allow imported subroutines to do this. In that code, the TOGGLE task from Figure 6.1 has been modified to work in a package that can be imported from a separate file. By using references, the task can use module variables.

The task of Figure 6.24 works as intended, toggling the module clock as many times as the input variable TIMES says to toggle it. The rate at which it toggles required a bit of trickery: the period is a parameter, but parameters cannot be referenced. To finesse this limitation, an integer variable was created, assigned the value of the parameter, and then referenced in the package.

This points out the limitation of using references. They work with variables but not parameters. The second subroutine of the package would be more useful if it were scalable rather than hard coded to eight bits. Since setting the size of scalable variables is the primary usage of parameters in synthesizable code, this limitation does reduce the utility of offloading subroutines to packages in separate files.

```
module import_subs(output logic P, N);
  parameter SIZE = 8;
  parameter PERIOD = 10;
  //convert parameter PERIOD to a variable so it can be referenced
  int PER = PERIOD;
  reg CLK;
  logic [SIZE - 1 : 0] DATA;
  import subs::*;
  initial TOGGLE(CLK, PER, 12);
  always_ff @(posedge CLK) par(DATA, P, N);
  initial begin
    DATA = 0;
    #PERIOD DATA = 1;
    #PERIOD DATA = 2;
    #PERIOD DATA = 3;
    #PERIOD DATA = 4;
    #PERIOD DATA = 5;
  end
endmodule


package subs;
task automatic TOGGLE(ref CLK, ref int PER, input integer TIMES);
    repeat (TIMES) begin
       #(PER / 2) CLK = 1'b0;
       #(PER / 2) CLK = 1'b1;
    end
endtask

//SIZE parameter cannot be referenced: a value is hard coded
//in the function, limiting its utility. A variable cannot be
//used to set the size of another variable, so the workaround
//used in task TOGGLE will not work here.
function automatic void par(ref [7:0] DATA, output logic P, N);
    P = ^DATA;
    N = ~^DATA;
endfunction
endpackage
```

■ FIGURE 6.24 Using references in packaged subroutines

Since this example uses initial blocks, one of which is used to create a clock signal, it would not be synthesizable. The clock generator task would itself be meaningless for synthesis, although the parity generator function could be used in a circuit design.

INTERFACES

SystemVerilog interfaces are a construct for grouping together related signals. An interface is a design entity like a module, although with its own rules and protocols. Like modules, as many instances of an interface can be created as needed. Interfaces can not only be scaled with parameters, but through the use of "modports," they can also be customized for different locations in a design hierarchy.

Grouping-related signals together form a new hierarchical entity sounds a lot like structures, which were introduced in Chapter 3. There is indeed some overlap between what can be done with interfaces and structures, but interfaces, as their name implies, are intended to encapsulate the connections and communications between modules. While structures can be used for that to a limited extent, interfaces allow port direction to be taken into account and can include logical constructs, which are capabilities that structures lack.

Interfaces are hierarchical entities that can contain instances of other interfaces, parameters, variables, constants, processes, subroutines, and continuous assignments. They cannot contain instances of modules or primitive operators.

In Figure 6.25, the flipflop-based register file design from Figure 3.17 is modified to use an interface. First an interface is created. It may be in a separate file that is compiled with the other design files. Then the interface is instantiated in the port list of the module. In the example, the name of the interface is BUS and it is given the instance name of ADBUS, for Address/Data bus, when instantiated. Interfaces still must follow the rule for multiply driven signals, so the data bus is declared to be a wire, which is a net type. When referencing interface signals, hierarchical names must

```
interface BUS; //interface and endinterface are SV keywords
  logic [7:0] ADDR;
  wire [7:0] DATA; //bidirectional signal must be a net type
  logic OE, WS;
endinterface

module REGFILE (BUS ADBUS); //ADBUS is an instance of interface BUS
  reg [7:0] MEM [0:255];
  reg [7:0] DATA_REG;

  assign ADBUS.DATA = ADBUS.OE ? DATA_REG : 8'bz;

  always_comb
  DATA_REG = MEM[ADBUS.ADDR];

  always_ff @(posedge ADBUS.WS) MEM[ADBUS.ADDR] <= ADBUS.DATA;
endmodule
```

■ **FIGURE 6.25** Register file with an interface

be used, as is done in the example with signals ADBUS.DATA, etc. Leaving out the instance name of the interface would be an error, as the module does not have local copies of the bus signals.

Like modules, interfaces can benefit from being scalable. Rather than hard coding the widths of the address and data buses, parameters may be used. The example of Figure 6.25 is reworked in Figure 6.26 to use parameters in both the module and interface. Although the interface is instantiated in the module port list, the parameters in the module do not push down to change the values of those in the interface. Both interface and module need to be instantiated in a top-level design to have one place where all parameters may be changed for the hierarchy. This is done in Figure 6.26 via the addition of module TOP, where both the interface and module are instantiated. In that example, the width of four and depth of six will be pushed down to both the interface and the register file. It would be possible to instantiate TOP in a still higher-level design and push new parameter values down from there throughout the hierarchy. Since the default values for the parameters in the interface and register file differ, changing at

```
interface PBUS #(parameter WIDTH = 8, DEPTH = 8);
  logic [DEPTH - 1:0] ADDR;
  wire [WIDTH - 1:0] DATA;
  logic OE, WS;
endinterface

module PREGFILE(PBUS ADBUS);
  parameter WIDTH = 16;
  parameter DEPTH = 16;
  reg [WIDTH - 1 : 0] MEM [0:2**DEPTH - 1];
  reg [WIDTH - 1 : 0] DATA_REG;

  assign ADBUS.DATA = ADBUS.OE ? DATA_REG : 'bz;

  always_comb DATA_REG = MEM[ADBUS.ADDR];

  always_ff @(posedge ADBUS.WS) MEM[ADBUS.ADDR] <= ADBUS.DATA;
endmodule

module TOP(input [7:0] ADDR, inout [5:0] DATA, input OE, WS);
  parameter WD = 4, DP = 6;
  PBUS #(.WIDTH(WD), .DEPTH(DP)) ADBUS(.*);
  PREGFILE #(.WIDTH(WD), .DEPTH(DP)) MEM(ADBUS);
endmodule
```

■ **FIGURE 6.26** Scalable register file and interface in a top-level design module

least one of them from above will likely be necessary for proper circuit functioning.

It might seem logical to just eliminate the parameters from the interface and have the interface automatically use the parameters from the instantiating module, PREGFILE in Figure 6.21. This approach works with subroutines. However, it does not work with interfaces. As in modules, parameters must be declared in the entity in which they are referenced.

Another unsuccessful idea to avoid use of a top-level module would be to use the defparam command in module PREGFILE to change the parameter values of instance ADBUS. This too is currently unsupported and attempts to use it crash both simulators and synthesizers.

In the examples shown, use of interfaces does not offer any obvious advantages. The interface must be created and instantiated. It is another file to be maintained. It forces use of hierarchical names where none were needed before. It complicates the use of parameters. Instantiated once, an interface is additional work without gain. The power of interfaces comes from reuse.

A generic computer architecture is shown in Figure 6.27. Six functional units are connected together via a common bus. By instantiating one interface in each functional unit rather than replicating each signal individually six times, not only is the coding of each module simplified, but design maintenance and reuse capability are also enhanced. A change to the interface need be made only in the interface, not in every module that instantiates it.

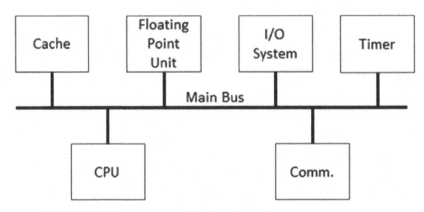

■ **FIGURE 6.27** Generic computer architecture with a common bus

Since each of the blocks is also likely to be a hierarchical design, the interface may be instantiated dozens of times. Big designs can have hundreds of ports, with the port list running on for pages. Such a structure offers many opportunities for errors to creep in. Instantiating one interface cleans up the design, makes it easier to understand and maintain, reducing the probability of errors. The bigger and more complex the design, the more benefit there is to be derived from the use of interfaces.

INTERFACE MODPORTS

When an interface is instantiated in multiple modules, there are likely to be some differences in how individual signals of the interface are needed in the various modules. Signals that are inputs to one will be outputs from another and vice versa. Also, some signals will be generated by the modules while others, such as a primary clock, will be sourced from the outside world.

In order for a synthesizer to be able to create the appropriate drivers for the signals of an interface, circuit description code must include additional information on direction. These data are specified in "modports." Simulation of interfaces in behavioral code without modports may indicate correct functioning, but that is not circuit design. A synthesizable circuit description needs modport direction data.

Adding modports to an interface represents something of a compromise from the ideal of interfaces, which is to have one common structure that need only be instantiated in each device. With modports, there still may only be one interface, but it will have different subtypes, complicating the interface and its maintenance. Nevertheless, use of interfaces is highly recommended, as their benefits still outweigh the additional complication of adding modports.

In Figure 6.28, the register file interface has been modified to include two modports, one for the register file itself and one for the processing unit that will source the address bus, output enable, and write strobe. The register file then must instantiate not the interface itself, but a modport of the interface. With signal directions specified in the modports, synthesizers will be able to infer the correct drivers for both the bidirectional data bus and the unidirectional signals driven by the processor.

The modports of Figure 6.28 are fine for a situation where all signals are internal to the device being designed, but frequently an

```
interface PBUS_modport #(parameter WIDTH = 8, DEPTH = 8);
  logic [DEPTH - 1:0] ADDR;
  wire [WIDTH - 1:0] DATA;
  logic OE, WS;
  modport MEMSIDE(input ADDR, OE, WS, inout DATA);
  modport PROCSIDE(output ADDR, OE, WS, inout DATA);
endinterface

module PREGFILE_modport(PBUS_modport.MEMSIDE ADBUS);
  parameter WIDTH = 16;
  parameter DEPTH = 16;
  reg [WIDTH - 1 : 0] MEM [0:2**DEPTH - 1];
  reg [WIDTH - 1 : 0] DATA_REG;

  assign ADBUS.DATA = ADBUS.OE ? DATA_REG : 'bz;

  always_comb DATA_REG = MEM[ADBUS.ADDR];

  always_ff @(posedge ADBUS.WS) MEM[ADBUS.ADDR] <= ADBUS.DATA;
endmodule
```

■ **FIGURE 6.28** Register file with an interface using modports

interface will also have primary input signals. In the example of Figure 6.29, the interface has externally sourced clock and reset signals as well as internal logic signals. In this example, clock and reset primary inputs go to all modules, the bus master sources

```
interface CLOCKEDBUS(input CLK, RST);
  wire [7:0] DATA;
  logic [3:0] STATE;
  logic ENABLE, READ, WRITE;
  modport MASTER(input CLK, RST, output STATE, ENABLE, READ,
    WRITE, inout DATA);
  modport SLAVE(input CLK, RST, STATE, ENABLE, READ, WRITE,
    inout DATA);
endinterface : CLOCKEDBUS
```

■ **FIGURE 6.29** Interface with primary inputs

internal signals STATE, ENABLE, READ, and WRITE and there is a bidirectional data bus that both master and slave can drive. All these signals are encapsulated in the interface.

Simulators have no concept as to which files are circuit design and which are for verification purposes only, so a hierarchy with clock and reset sourced from a test program and declared simply as signals in an interface may simulate correctly. However, for synthesis it will be necessary to make clock and reset signal inputs to the interface.

The interface CLOCKEDBUS shown in Figure 6.29 ends with an optional repetition of the interface name preceded by a colon. While seemingly redundant, use of this feature helps readers navigate designs as they get more complicated.

Although instantiations of modules and primitives are forbidden in interfaces, nearly anything else goes, including functional blocks. This can prove to be useful, for example, in a situation where a sequential handshake protocol would otherwise need to be replicated in each module that uses the interface.

In the example of Figure 6.30, there is an interface that includes logic for a handshake protocol. One side sources a VALID signal and the other a READY. When instantiated, each modport will include only the flipflop that it drives. In this example, no code is included to drive the bidirectional data bus. Designs DRIVER and RECEIVER contain no logic beyond that contained in their interface modport. This simplification was done to emphasize the procedures for instantiating the modports with their logic. If the code were to be synthesized, everything related to the DATA port would be optimized out, as it has no drivers.

To be useful, designs DRIVER and RECEIVER would have to include code beyond that included in interface HANDSHAKE_INT. To prevent the synthesizer from optimizing out all the logic, signals VALID and READY were made into primary outputs and are thus in the interface's port list. In a real system, they likely would be internal signals only.

```
interface HANDSHAKE_INT(input CLK, RST, output logic VALID, READY);
  wire [7:0] DATA;
  modport SLAVE(input CLK, RST, READY, output VALID, inout DATA);
  modport MASTER(input CLK, RST, VALID, output READY, inout DATA);

  always_ff @(posedge CLK, negedge RST)
    if (!RST) VALID <= 1'b0;
    else
      if (READY) VALID <= 1'b1;
      else VALID <= 1'b0;

  always_ff @(posedge CLK, negedge RST)
    if (!RST) READY <= 1'b1;
    else
      if (VALID) READY <= 1'b0;
      else if (READY) READY <= 1'b1;
      else READY <= READY;
endinterface : HANDSHAKE_INT

module DRIVER(HANDSHAKE_INT.MASTER  I);
endmodule : DRIVER
module RECEIVER(HANDSHAKE_INT.SLAVE J);
endmodule : RECEIVER

module TOP(input CLK, RST, output VALID, READY);
  HANDSHAKE_INT H(CLK, RST, VALID, READY);
  DRIVER D(H.MASTER);
  RECEIVER R(H.SLAVE);
endmodule : TOP
```

■ **FIGURE 6.30** Interface with behavioral code implementing handshake protocol

■ **SUMMARY**

Tasks and functions are useful constructs for offloading commonly used and/or complex routines into reusable coding structures that can be maintained independently from the modules in which they are used. These subroutine constructs are found more often in test fixtures than in circuit description, but they are useful in both. Judicious use of subroutines makes modules easier to understand and maintain.

Because functions need to run in zero simulation time, they can only be used to represent combinational circuitry. Tasks can have inferred delays such as waiting for a clock edge, although they cannot include functional blocks. When it comes to hardware design, tasks too are generally restricted to representing combinational logic. From the restriction on timing statements in functions, it follows that tasks can call functions but functions cannot enable tasks.

Interfaces are a method of grouping together related signals and protocols. As with subroutines, use of interfaces can simplify modules. Instead of replicating commonly used signals and data transfer protocol in multiple modules, a single interface can be created and instantiated multiple times. Interfaces can be customized through the use of modports to allow one standard interface to be configured for different environments.

Subroutines and interfaces are powerful tools for making circuit descriptions shorter, more readable, and more reusable. Although there is not any design that is impossible to do without them, their use is highly encouraged in all professional environments.

Synchronization

The previous chapters have covered the parts of Verilog and System-Verilog that are useful for circuit design. Using them to create reliable circuits, however, requires more expertise. Lack of understanding of asynchronous interfaces is one of the most common sources of failures in digital circuits.

This chapter examines the theoretical basis for circuit failures due to timing errors and supplies a set of solutions that can be applied to different types of synchronizing challenges.

LATCH INSTABILITY

One of the primary reasons for unreliable operation in digital circuits is improper design of asynchronous interfaces. This section will cover the various categories of asynchronous interfaces and the techniques for making them as reliable as possible.

While all circuits worth considering are clocked, not all events and signals are synchronous with the clock. It is the arrival of an asynchronous event at a clocked flipflop that causes most transient circuit malfunctions. There is no way to totally eliminate any possibility of such events, but there are ways to minimize the probability that they will cause systemic failures.

The root of all stability problems is the potential instability of an RS latch when presented with illegal input conditions. For an RS latch made from NAND gates, having both inputs equal to zero is the illegal condition and for a NOR RS latch, it is having both inputs equal to one. Not only are the outputs unpredictable for such input conditions, but in the worst case they can cause oscillation, ultimately leading to physical circuit failure, not just transient errors.

The two fundamental types of RS latches are shown in Figure 7.1. Truth tables for their behavior are shown in Table 7.1. Variations

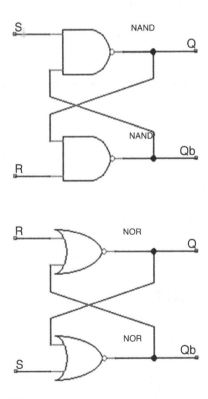

■ **FIGURE 7.1** RS latches

Table 7.1 RS latch functions

NAND Latch

R	S	Q	Qb
0	0	Unstable	Unstable
0	1	0	1
1	0	1	0
1	1	Hold State	Hold State

NOR Latch

R	S	Q	Qb
0	0	Hold State	Hold State
0	1	1	0
1	0	0	1
1	1	Unstable	Unstable

can have more inputs, but the problem of illegal input conditions remains no matter how many inputs each gate has.

If the latches of Figure 7.1 are simulated, the problem is not apparent. This demonstrates a limitation of digital simulation, not the absence of problems. The issue is sequential evaluation of parallel events. When the latches are simulated, first one gate output is calculated and then the other. While there is no guaranteed order of evaluation that one will be evaluated first and then the results of that evaluation applied to the second is a certainty. This is not how the physical circuit would work, where both gates would be transitioning simultaneously.

In a digital simulation of a NAND latch, application of the illegal input conditions of R = 0 and S = 0 results in both outputs set to logic one. While improper operation for supposedly complementary outputs, the outputs are stable and the results repeatable. There is no indication that circuit failure can result. However, with physical, silicon gates, the two outputs will be simultaneously stimulated and the transitioning outputs will be inputs to the other gate. Possible results include oscillation and having one or both outputs staying in an illegal, intermediate level, neither logic one nor logic zero. Since such an intermediate level would also result in the gate transistors operating in the high power dissipation linear zone, both oscillation and indeterminate logic level can result in permanent physical damage to the circuit, as well as the more likely result of transient incorrect logical operation.

While a NAND or NOR gate without feedback will operate according to its Boolean function regardless of the relationship between input signal arrival times, when feedback is added, as it is in RS latches, timing matters. When both inputs simultaneously transition to 00 for a NAND latch or 11 for a NOR latch, circuit failure is a real possibility.

This potential for improper operation is the heart of all synchronization problems.

FLIPFLOPS, LATCHES, AND VIOLATIONS

A gate-level model of a flipflop is shown in Figure 7.2. In that model, it is apparent that flipflops can be little more than RS latches connected in series, although there are many flipflop designs in which the latches are not so clearly differentiated. Unlike a standalone RS latch, where the only way to have simultaneously switching inputs is to apply such inputs simultaneously to the R and S inputs, in a flipflop there are several different paths from the circuit inputs to the latches. Even if no two signals arrive at any two primary inputs simultaneously, there may still be simultaneous arrivals at one of the latches.

If the flipflop of Figure 7.2 does need to be put into a known initial state at reset time, the input can be gated with a reset signal as shown in Figure 7.3 [1]. Such an approach depends on having both edges of the reset signal already synchronous with the master clock, having the clock running when the reset arrives and having the reset signal be active long enough that it can always be synchronized with the clock. Since those conditions are often not met, a synchronous reset may not be an option. A further disadvantage of such an approach is that there will need to be an additional gate delay in all data paths going to flipflops that need to

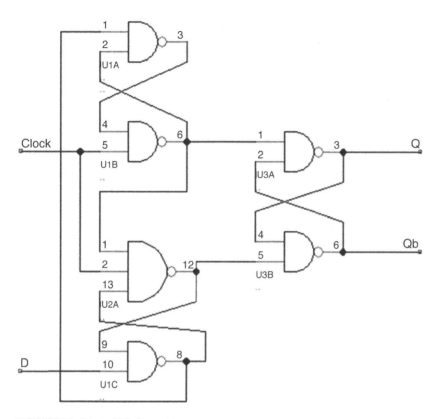

■ **FIGURE 7.2** Gate level D flipflop model

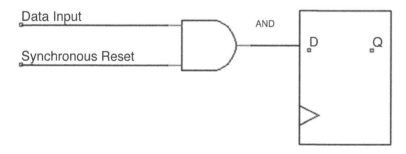

■ **FIGURE 7.3** Synchronously resetting a flipflop with an active-low reset

be reset, possibly increasing both area and critical path. However, flipflops with asynchronous reset inputs may be larger and slower than those without, so there is not necessarily a performance penalty for implementing a synchronous reset.

SystemVerilog code inferring a flipflop with a synchronous reset is shown below. The only difference between it and code for a flipflop with an asynchronous reset is in the sensitivity list. A synchronous reset flipflop has only the clock and not the reset there.

```
//Synchronous reset flipflop
always_ff @(posedge CLK)
  if (!RST) Q <= 1'b0;
  else Q <= D;
```

In Figure 7.4, the flipflop shown in Figure 7.2 was modified to add an active-low asynchronous reset input. While assertion of this signal will override whatever the clock and D inputs are doing,

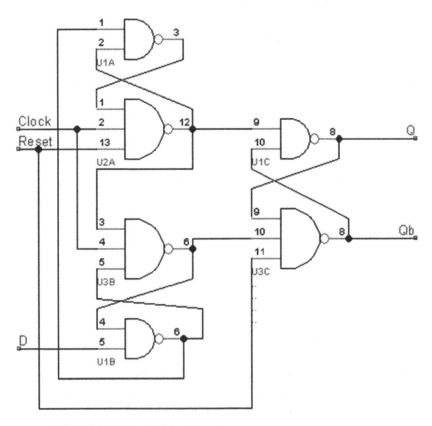

■ **FIGURE 7.4** Gate level D flipflop with reset

the diagram also shows that reset too is an input to internal RS latches and that its arrival at those latches may be coincident with that of the clock even if the two inputs do not transition simultaneously. This may be another source of timing problems.

In order to prevent unpredictable and damaging operation, each flipflop will have operating parameters that specify the relationships between arriving signals. The most common of these parameters are setup time, hold time, and recovery time, which are shown in Figure 7.5. These operating parameters specify what relationships must be maintained between signal arrival times. For setup time, there will be some amount of time before the active edge of the clock that the data input must be stable. Hold time is the amount of time after the clock edge that the data input must remain stable. Release time specifies how long there must be between deassertion of reset and the arrival of the active edge of the clock. Violations of these operating parameters are the primary causes of circuits malfunctioning.

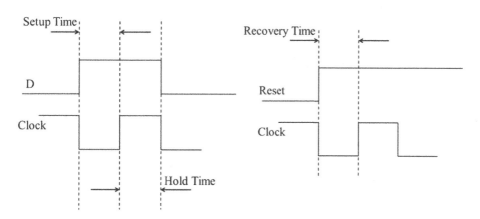

■ **FIGURE 7.5** Flipflop operating limitations

When a signal enters a sequential circuit asynchronously, avoiding any violation of the operating limits can be impossible, but steps can be taken to reduce the probability of damage to an infinitesimal figure. One common case is an asynchronous reset. Registers in control blocks almost always need some mechanism for being put into a known initial state, and this is usually done via a reset signal.

Behavioral code for a D flipflop with an asynchronous reset was introduced in Figure 5.18 and is reproduced in Figure 7.6. If the active-low reset is asserted, the proper behavior of the code and any circuit synthesized from it is to set the output to logic zero, regardless of clock status and regardless of whether the clock is running or not. It should remain logic zero as long as reset is asserted and resume clocked operation as soon as reset is deasserted. The desired behavior is shown in Figure 7.7, where the output is set to zero following assertion of the reset signal. Both clock and data inputs are don't care. The Q output starts out as unknown but goes to logic zero following the assertion of RST. Both unknown and don't care are shown as X values.

```
module flipflop(input CLK, RST, D, output logic Q);
  always_ff @(posedge CLK, negedge RST)
    if (!RST) Q <= 0;
    else Q <= D;
endmodule
```

■ **FIGURE 7.6** Behavioral code for a D-type flipflop with asynchronous reset

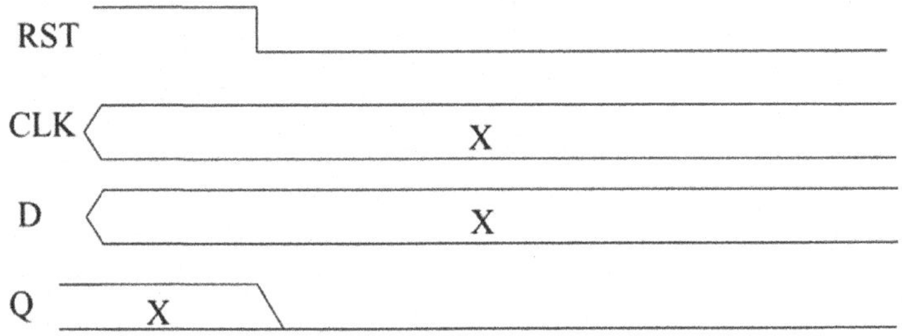

■ **FIGURE 7.7** Output is set to logic zero following assertion of reset

In behavioral simulation, that is what will happen. In a physical circuit, it may not.

The problem is that the reset does not stay active forever, although if it did, the circuit would never do anything useful. At some point, reset is released and the flipflop starts operating synchronously with the clock. If nothing is done to prevent it, the release of reset can come at an arbitrary time in relation to the clock phase, which may result in a circuit failure. This is illustrated in Figure 7.8, where the rising edge of RST is coincident with the rising edge of CLK. When this results in the two edges arriving simultaneously at one of the RS latches in the flipflop, a recovery violation occurs and damaging instability may result. Instead of being put into a known state to start normal operation, the output is unknown and may not be a stable, legal logic level at all.

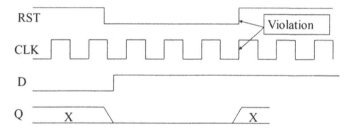

■ **FIGURE 7.8** Recovery violation

ASYNCHRONOUS ASSERT, SYNCHRONOUS DEASSERT

While reset must be capable of putting the registers into a known state regardless of clock operation or any other concurrent circuit operation, to avoid a recovery violation it must be deasserted synchronously with the clock. This seemingly contradictory set of requirements is easily met by adding an asynchronous assert, synchronous deassert (AASD) module between the asynchronous reset input and the internal reset distribution network. Code for the AASD system is shown in Figure 7.9, and simulation of the operation of the AASD system is shown in Figure 7.10.

As soon as the reset signal is asserted, the module output is also asserted. This is true despite the fact that the clock is not operating, that it has not even been initialized. Assertion is thus asynchronous. Eventually the clock does start up, but the AASD

```
module aasd(input CLK, RST, output logic AASD_RST);
  reg STAGE1;
  always_ff @(posedge CLK, negedge RST)
    if (!RST) begin
      STAGE1 <= 1'b0;
      AASD_RST <= 1'b0;
    end
  else begin
    STAGE1 <= 1'b1;
    AASD_RST <= STAGE1;
  end
endmodule
```

■ **FIGURE 7.9** AASD circuit description

■ **FIGURE 7.10** AASD operation

output remains stable and low until after reset is released. Once reset is released, the flipflops resume normal operation. On the rising edge of the clock following the release, the internal flipflop STAGE1 is set to logic one. On the next rising clock edge, the logic one on STAGE1's output is propagated to the AASD output flipflop, which can then be distributed to the entire reset network of the design. The release of reset will always lag the clock, avoiding any possibility of a recovery violation at subsequent flipflops. Thus, this module meets the desired standard of AASD.

In the simulation shown in Figure 7.10, RST is released midway between rising clock edges. Because it is asynchronous with respect to the clock, it could come at any time, possibly leading to a recovery violation in the AASD module. If that were to happen, the output of flipflop STAGE1 could be unstable. This is where the probabilistic nature of asynchronous design solutions comes into effect.

In theory, a recovery violation on STAGE1 could propagate to the AASD output flipflop, which could then propagate it to the rest of the circuit. For that to happen the illegal output condition would have to persist for a full-clock period and then the illegal output on the second flipflop would also have to persist until it interferes with operation at some other register in the circuit.

That is extraordinarily unlikely to happen because the probability of an output remaining in an illegal state decays exponentially with time. As will be shown mathematically later in this chapter, the more flipflops that are strung together before the synchronized reset is distributed, the lower the probability of an illegal condition persisting to the point where it causes any damage. Just two flipflops are almost always enough.

Using an AASD module will solve the synchronization problem for starting normal operation after a reset, but an external reset input is far from the only asynchronous signal a digital design will typically have to deal with. Embedded systems will have asynchronous signals coming in from sensors, personal electronics will have keyboards and other pushbutton inputs, networked computers will have asynchronous communications links: All are examples of data that will need to be synchronized before they can be used.

SLOW-SPEED SINGLE-BIT CLOCKED ASYNCHRONOUS INTERFACES

A more general case of an asynchronous input signal is illustrated in Figure 7.11. In that figure, a signal generated in one system, synchronous with its own clock, needs to be read into a system with a different clock. Because there is no fixed phase relationship between the two clocks, sometimes the signal will be read in correctly and sometimes it will transition too close to an edge of the receiving clock, resulting in a setup or hold violation and an indeterminate signal value. There is a high probability that Data 2 will eventually resolve to either a logic one or a logic zero, but there is no way to predict which value it will settle on.

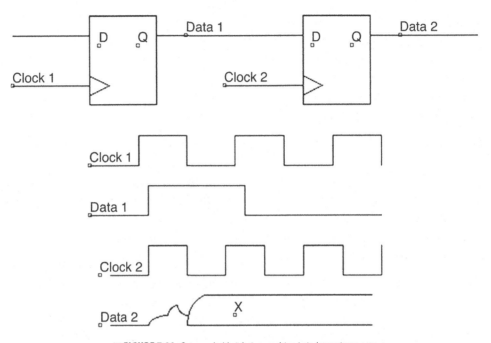

■ FIGURE 7.11 Setup or hold violation resulting in indeterminate output

In a situation where the input signal changes slowly in relation to the clock, it may be good enough simply to ensure that unstable, indeterminate signals are not propagated beyond the asynchronous interface. In the example of Figure 7.12, a signal needs to be read by a digital system, but the signal moves slowly. It stays in each state for many clock cycles.

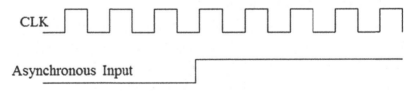

■ FIGURE 7.12 A slow signal being read by a clocked digital circuit

If the asynchronous signal is run through two flipflops before being used as a logic signal, just as was done with the AASD module, the probability of an illegal state propagating to a place where it will do any damage asymptotically approaches zero. Code for a synchronizer module and the hardware it would produce are shown in Figure 7.13. DATA_IN is an asynchronous input and DATA_OUT is synchronized with the clock. It will typically have a vanishingly small, although still theoretically nonzero, probability of being anything other than a logic zero or logic one.

```
module synchronizer(input CLK, DATA_IN, output logic DATA_OUT);
  logic STAGE1;
  always_ff begin @(posedge CLK)
    STAGE1 <= DATA_IN;
    DATA_OUT <= STAGE1;
  end
endmodule
```

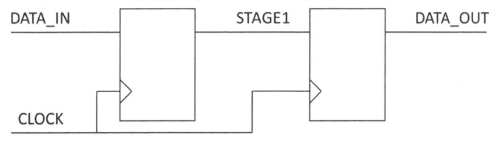

■ **FIGURE 7.13** Code for a two-stage synchronizer and resulting circuit

HIGH-SPEED SINGLE-BIT CLOCKED ASYNCHRONOUS INTERFACES

This simple synchronizer will be satisfactory for situations where the input signal is slow relative to the clock because eventually the new value will be recognized and any unstable, illegal states will be confined to the synchronizer.

The need for a synchronizer between signals generated in different systems is constant even if both are clocked at nominally the same frequency.

The overwhelming majority of digital systems have a quartz crystal providing a fundamental time base, although there are other options with more or less frequency precision. Regardless of which oscillator technology is used and what the nominal frequency is, no two will operate at precisely the same frequency to an infinite number of decimal places, yet that unobtainable situation would be necessary for two systems to operate together without synchronizers.

A crystal will typically have a tolerance of single-digit to tens of parts per million (PPM), varying with temperature. For illustrative purposes, assume an overly optimistic situation with two 10 MHz crystals each operating with 1 PPM tolerance. Further assume that one is fast by one PPM and the other slow by the same amount, that is, one is operating at 10,000,010 Hz and the other at 9,999,990 Hz.

With these operating parameters, the two systems would be out of sync by 20 clock cycles after a single second of operation. Exchanging data between the two would be utterly unreliable, as sometimes there would be approximately a clock cycle between data generation and reception and other times the two would be simultaneous. The phase relationship between the two sides would be continuously changing, an impossible situation for reliable operation. Even with two systems working at nominally the same frequency to a precision unobtainable under most circumstances, data transfers require synchronization.

Fast and/or parallel data transfers require more complex solutions, which also will be presented in this chapter, but a bit serial slow interface can be made to work with nothing but a two-stage synchronizer, as shown in Figure 7.13.

The question remains, however, why two flipflops will be satisfactory. Why not one or three or something else entirely?

The phenomenon of a logic signal outputting an illegal, indeterminate value is called metastability. Using long-obsolescent transistor–transistor logic (TTL) values for an example, a legal TTL logic zero input can range from zero to 0.8 V. A legal TTL logic one input can range from 2.0 to 5 V. If an input is between 0.8 and 2.0 V, it is neither a logic one nor a logic zero, but an illegal, indeterminate state.

To transition from one legal state to the other, a signal must momentarily pass through the illegal values. It is when a signal fails to complete the transition, and it gets stuck somewhere in the middle, that it enters into a metastable state. There is no theoretical limit as to how long a signal may stay there, although like a ball balanced on a knife-edge, it usually does not stay long before falling off to one of the legal logic levels.

The root cause of a signal getting into a metastable state is always a violation of the operating parameters of a flipflop. In a synchronous system, avoiding such violations is straightforward. As long as the sum of the clock to Q time of the driving flipflop, the combinational and wiring delay between flipflops and the setup time of the receiving flipflop are less than the period, there will not be any violations. For the typical synchronous design as shown in Figure 7.14, with a flipflop propagation delay (Clock to Q) of t_{cq}, combinational delay of t_{logic} and flipflop setup time of t_{su}, the total path is

$$\text{Path} = t_{cq} + t_{logic} + t_{su}$$

There is no metastability risk as long as the clock period is greater than the path.

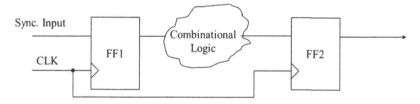

■ **FIGURE 7.14**   No risk of metastability in a synchronous system when path is less than period

When the input is asynchronous, there is an additional factor. If the period is not long enough for a metastable output to resolve to a legal logic level in addition to all the other path factors, the metastable error will propagate. Figure 7.15 is identical to Figure 7.14 except that the input is now asynchronous. This system will only work if the resolution time t_r is less than the period t_p

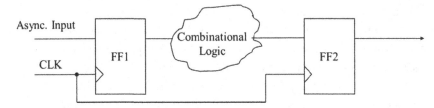

■ **FIGURE 7.15** With an asynchronous input, metastability resolution time is added to the path

minus all the other path factors as defined in the path equation above. The resolution inequality is shown below.

$$t_r \leq t_p - t_{cq} - t_{logic} - t_{su}$$

Because there is no fixed relationship between the asynchronous signal's arrival time and the clock, sometimes the new data will be read in correctly and sometimes they will not be. There will always be some risk in a system like this of metastability errors, but as already noted, the probability of staying in a metastable state long enough to cause a systematic failure decays exponentially with time. The key to reliable asynchronous interface design is reducing that probability to a minuscule and acceptable level.

In the resolution inequality, each element of the inequality is an independent variable. Each can be adjusted to achieve a satisfactory design. The only part that cannot be changed is that there is an asynchronous input coming into a clocked design.

Just as a ball can theoretically stay balanced on a knife-edge forever, the possibility of failure will never go to zero, but the circuit can be changed to reduce the probability of damaging metastability.

The options are as follows:

1. Decrease the operating frequency. As t_p goes up, so does the time available for t_r.
2. Eliminate combinational logic. Take t_{logic} out of the equation entirely.
3. Use different flipflops, ones operate faster and have greater resistance to metastability.

A flipflop risks outputting an illegal, metastable state when one of its operating limitations is violated, as shown in Figure 7.5. Such violations are inevitable with an asynchronous input. However, the illegal state will only cause a circuit malfunction if it persists long enough to propagate elsewhere in the circuit. If the metastable flipflop reverts to a legal state before its signal is clocked in to the next flipflop, there will be no failure.

While it is impossible to predict how long a flipflop will remain in a metastable state for any individual occurrence, it is known that the probability of a flipflop remaining in such a state decays exponentially over time according to the resolution function equation

$$f_{\rm r} = e^{(-t_{\rm r}/\tau)}$$

where $f_{\rm r}$ is the probability that the flipflop output will remain in an illegal state, $t_{\rm r}$ is the resolution time introduced in the resolution inequality above, and τ is a time constant, a parameter of a specific flipflop.

Resolution time is not a flipflop parameter, although flipflop parameters are components of it. It is the amount of time available in a specific circuit for the flipflop to resolve to a legal value without propagating an error.

In Figure 7.16, $t_{\rm r1}$, the resolution time of the top path, is shorter than $t_{\rm r2}$, the resolution time of the bottom path. The bottom path has the entire period (minus the receiving flipflop's setup time) for the output of the driving flipflop to turn into a legal value, whereas in the top path the flipflop output must be at a legal value with enough time to spare to have it propagate through the combinational logic cloud and still arrive at the receiving flipflop, a setup time before the clock edge.

The standard measurement for circuit reliability is "mean time between failures," or MTBF, which is the inverse of the failure rate. For an asynchronous event entering a circuit as shown for

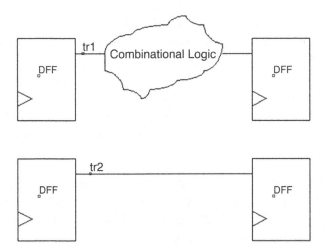

■ **FIGURE 7.16** Resolution time as a function of circuit design

either path of Figure 7.16, the MTBF can be calculated from the following equation:

$$\text{MTBF} = \frac{e^{(t_r/\tau)}}{t_0 f_c f_d}$$

In the MTBF equation, resolution time is again the amount of time available in a circuit for any metastability to resolve to a legal logic level without causing any damage. Note that in the resolution function equation, the exponent is negative, leading to an asymptotic approach to zero as time approaches infinity. Conversely, in the MTBF equation, the time between failures grows with resolution time: the more time available for the ball to fall off the knife-edge, the longer the time between failures.

Both τ and t_0 are time constants of the flipflop, which may be obtained from the semiconductor manufacturer or derived from experimental data. f_c is the clock frequency and f_d is the rate at which the asynchronous data are arriving. These are operating conditions of the circuit, not semiconductor parameters.

In Figure 7.17, asynchronous data are read into a clocked system. If the second flipflop outputs an illegal value, it can propagate

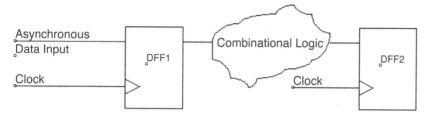

FIGURE 7.17 MTBF demonstration circuit

throughout the system, resulting in a failure. Given flipflop and circuit operational parameters, the MTBF can be calculated and a decision made as to whether or not it is satisfactory.

Texas Instruments conducted some investigations into the performance of their 7400ABT series components, deriving from experimental data and SPICE models values of 16.9 ps for t_0 and 330 ps for τ [2]. Using these characteristics, a circuit similar to the one shown in Figure 7.17 was proposed. In their demonstration, the worst path through the combinational logic cloud was set to 9 ns and the second flipflop had a setup time of 5 ns. The flipflops were driven at 33 MHz, giving a resolution time (t_r) of 16 ns. The data rate was set to 8 MHz.

All the necessary data were then available to calculate the failure rate:

$$t_0 = 16.9 \text{ ps}$$
$$\tau = 0.33 \text{ ns}$$
$$t_r = 16 \text{ ns}$$
$$f_c = 33 \text{ MHz}$$
$$f_d = 8 \text{ MHz}$$

Applying the MTBF equation, it turned out that this circuit would, on average, fail once every 2.55×10^{17} s, or there would be about eight billion years between failures. Even if tens of millions of these circuits were to be produced, this is likely to be a satisfactory standard of reliability.

Modest changes in operating conditions can have a dramatic impact on reliability. Increasing the data rate to 12 MHz and the receiving clock frequency to 50 MHz, the resolution time goes down to 6 ns (20 ns period − 9 ns combinational delay − 5 ns setup time). With f_c and f_d changed to 50 and 12 MHz, respectively, the MTBF drops to about 7766 s, or a bit over 2 h. Reliability went down by 13 orders of magnitude for a 50% increase in operating frequency.

To improve reliability, a second stage of synchronization before distribution of the signal can be added. With another flipflop, the previous MTBF would be multiplied by another factor of $e^{(t_r/\tau)}$. The two-stage MTBF equation is shown below.

$$\text{MTBF}\,2 = \frac{e^{(t_{r1}/\tau)}}{t_0 f_c f_d} \times e^{(t_{r2}/\tau)}$$

Adding a second synchronizing flipflop with all other factors unchanged from the circuit that fails every 2 h, the MTBF goes back up by 19 orders of magnitude. The dramatic increase is due in large part to the relatively long 15 ns t_r of the second stage (20 ns period − 5 ns setup time), which is facilitated by the absence of any combinational logic in the path from the previous flipflop. If that still produces an unsatisfactory period between failures, more stages can be added, each one multiplying the reliability by $e^{(t_r/\tau)}$.

A further improvement can be obtained by keeping all the combinational logic associated with the incoming asynchronous signal in the clock domain of that signal. This architecture is shown in Figure 7.18. By keeping all combinational logic out of synchronizing paths, both stages of the synchronizer will have

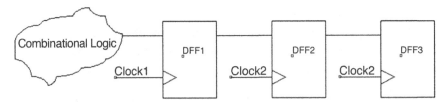

■ **FIGURE 7.18** Recommended asynchronous interface architecture

resolution times that are as long as possible, resulting in the most reliable asynchronous interfaces. This design style is always recommended.

Some semiconductor vendors have specialized metastable-hardened synchronizer cells that may be instantiated in clock domain crossings. If such cells are available, their use is preferable to using behavioral code to infer standard flipflops.

While the performance characteristics from the Texas Instruments study are from logic families long out of date and clock speeds have increased by around two orders of magnitude since those parts were state of the art, the design solutions have not changed. Even when operating near the practical speed limit for a logic family, two stages of synchronization are almost always enough for a single asynchronous interface.

So the answer to the question at the start of this section is that two flipflops are used because one synchronizing flipflop often is not enough and three are rarely necessary for adequately reliable operation.

MULTIPLE HIGH-SPEED SINGLE-BIT CLOCKED ASYNCHRONOUS INTERFACES

As the number of asynchronous interfaces in a design becomes large, the MTBF on any given interface may need to be far longer than would be necessary for any single-clock domain crossing, resulting in the necessity of adding more flipflops to some or all of the synchronizers.

Modern circuits may have many clock domains and even more data paths crossing clock domains. The probability of a circuit failure is the sum of all the individual probabilities of a failure on any one crossing. If there are 1000 clock domain crossings in a design, each one having the same MTBF, the overall MTBF will be reduced by three orders of magnitude. Therefore, each synchronizer may need to have more stages in order for the whole design to have a satisfactory level of reliability. The failure rate of a circuit with n clock domain crossings is the reciprocal of the sum of all the mean times between failures as shown in the equation below, where MTBFi is the MTBF for the ith clock domain crossing.

$$fn = \frac{1}{\sum_1^n \mathrm{MTBF}\, i}$$

ASYNCHRONOUS PARALLEL BUSES

A two-stage synchronizer can be added to an interface where only a single data bit is of concern and it changes slowly. However, that does not deal with all asynchronous interface challenges.

A multibit bus might also change slowly compared to the receiving clock. A two-stage synchronizer could then be added to each bit, as shown in Figure 7.19.

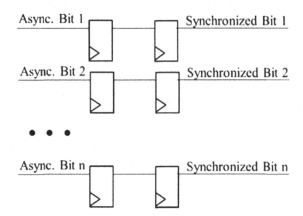

■ **FIGURE 7.19** Synchronizing flipflops added to each bit of a bus will lead to data incoherence

Such a scheme would not work. Even if every bit were precisely aligned upon arrival, already an impossible situation, there would still be a difference between rise and fall times. Usually rise times are slower than fall times. The problem this creates is illustrated in Figure 7.20. When the transmit clock (XCLK) rises, outputs start to change. The receive clock (RCLK) slightly lags XCLK in

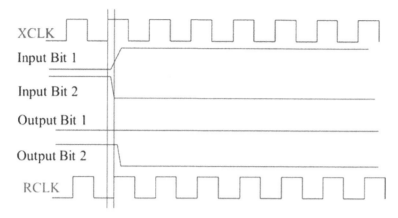

■ **FIGURE 7.20**   Rise times are typically longer than fall times

this example, although as they are asynchronous, the phase relationship between the two will be constantly changing.

As illustrated in Figure 7.20, the receive clock is delayed just enough from the transmit clock so that it reads all the falling bits as having completed the transition from logic one to logic zero and all the rising bits as still logic zeros. Thus, the word read will be all zeros, regardless of the value of the transmitted word. The received data word would be incoherent.

A modification of the two-stage synchronizer that will work under some circumstances for parallel buses is shown in Figure 7.21 [3]. In that design, an enable signal is synchronized with a two-stage synchronizer. Once that signal is synchronized with the receiving side, it is used to allow asynchronous data to be clocked in.

This scheme is limited to low data rates. If the clock on the sending side ever exceeds the rate on the receiving side, even for a data burst as small as two words, the system will fail. It also is subject to errors if the incoming data changes during the enable period. Since the synchronizer adds two clock cycles of latency to the enable signal that are not present on the data path, this arrangement can only be used when data stability over the entire protocol can be assured.

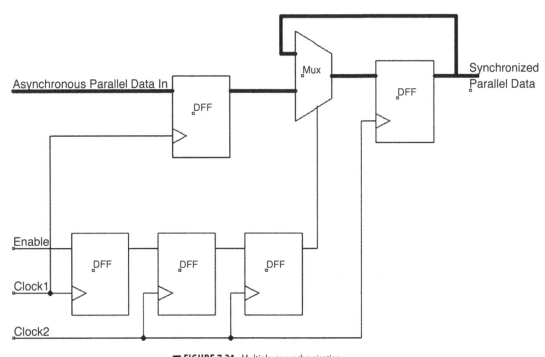

■ FIGURE 7.21 Multiplexor synchronization

FIFO

The more general solution for a design with an asynchronous bus input is to use a FIFO, a First In, First Out queue.

A common occurrence in digital communications is to have a burst of high-speed data that need to be processed over time. This is illustrated in Figure 7.22, where the Clock 2 timing domain sends a byte of data to Clock 1's domain, where it can be processed at a slower rate.

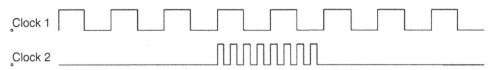

■ FIGURE 7.22 Data burst crossing clock domains

A situation like this cannot be handled by the multiplexor synchronizer illustrated in Figure 7.21. A FIFO is required.

A FIFO provides timing isolation between clock domains. Data may be sent into the FIFO with one clock and read out with a different, asynchronous one. A simple protocol is typically built in to each FIFO: if the FIFO is full, more data may not be sent. If the FIFO is empty, nothing may be read from it. Even with a FIFO, the long-term data rates must be identical. More data can never be put into one side than is read out from the other.

In a true FIFO, each data word passes through all unused stages of the FIFO until it either arrives at the output or is blocked by a full stage. This type of FIFO involves the use of cascaded latches and a nonoverlapping two-phase clock. The design is complicated and slow. It is not commonly used anymore. Instead, modern FIFO designs use a memory configured as a circular buffer.

A circular buffer FIFO consists of four subdesigns:

1. Memory, either a RAM or a register file
2. Read pointer
3. Write pointer
4. Control logic and empty/full flag generators.

The top level of a FIFO design of this type is shown in Figure 7.23 [4].

With such a design, the speed is nearly independent from the size, unlike a true FIFO, where speed is inversely proportional to depth. A circular buffer FIFO is as deep and wide as the memory cell it uses. The memory cell is also likely to be the component that limits the FIFO's operating speed, so the FIFO in all probability will be as fast as memory technology allows.

The FIFO model developed in this section will use a register file rather than a RAM cell. Although register files have the disadvantages of being larger and more power hungry than memory cells, they have the advantages of being synthesizable without any

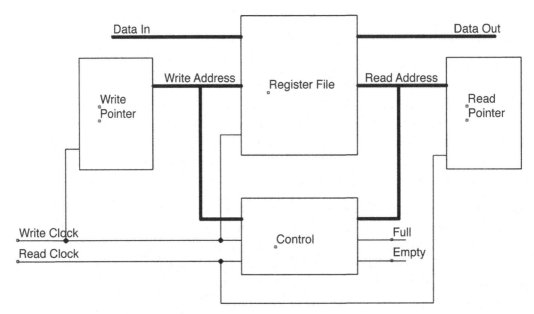

■ FIGURE 7.23 FIFO top-level block diagram

additional tools and matching perfectly to the needs of a FIFO. They also can be fast. Being made of flipflops, each cell naturally has an input port and an output port, exactly what is needed for a FIFO. While a dual-port memory is often used for FIFOs, a register file simplifies the design. The register file developed in previous chapters uses a single bidirectional data port, but that would not be suitable for a FIFO, as that would require arbitrating access to the port and would make simultaneous reads and writes impossible. Accordingly, the register file will be modified to have a data input port and a data output port.

In most commercial designs, only the smallest FIFOs would use a register file, as the power savings of using a memory cell would override the other concerns for larger applications.

The algorithm for FIFO operation consists of only three procedures:

1. Initialize to empty.
2. Allow write if not full.
3. Allow read if not empty.

Reading from an empty FIFO would be an error, as a random data word would be accepted as valid data. Writing to a full FIFO would also be an error, as valid data waiting in the queue would be overwritten by newer data. Reading from an empty FIFO is known as underflow. Writing to a full one is overflow. Both need to be avoided.

Because the objective of the FIFO is to bridge two clock domains, implementing this algorithm will require operations that use data from the two different domains.

The FIFO will write to and read from every memory location in sequence. This will require a write address and a read address, each pointing to locations in the same memory. Comparisons of the two pointers will be needed to determine if the FIFO is full, empty, or neither.

Comparison of the pointers is the most complicated part of the FIFO design, because one is generated from the read side clock and the other comes from the write side. If the design were synchronous, the comparison would be easy, but if the design were synchronous, the FIFO might not be needed at all.

Empty and full are determined by comparisons of the two pointers. Because the memory is used as a circular buffer, any value of the pointers can mean empty, full, or something in between.

The following sequence of events illustrates the inadequacy of simply comparing pointers to determine if the FIFO is full, empty, or neither empty nor full.

At initialization, the empty flag is set true and the full flag false. Both pointers are set to zero. Since the FIFO is empty, write is the only permissible operation.

After the first write, the write pointer will be at one and the read pointer still at zero. If a read then follows, both pointers will be at one. The FIFO will be empty again and the empty flag should be set.

Then there are N writes for an N deep FIFO without any reads. The write pointer will wrap around and will again be at one. The FIFO will be full, so the full flag should be set.

In the absence of any other information, having the pointers equal thus means both full and empty. Since it is not possible for a FIFO to be simultaneously full and empty, something else needs to be added to the design.

So a match between read and write pointers sometimes means the FIFO is empty and sometimes means that it is full. If it is a read operation that causes the pointers to match, then the FIFO is empty. If the write pointer catches up to the read as a result of a write operation, the FIFO is full. Simply comparing without knowing which one changed last is insufficient to determine if the FIFO is empty or full.

Expanding both pointers by one bit can solve this problem. If each pointer has an extra bit and if all bits are equal, then the FIFO is empty. If the most significant bits are opposite but all others are equal, then the FIFO is full.

A further complication is that the two pointers cannot be directly compared, as they are asynchronous. One is a function of the write clock, the other of the read clock. However, compared they must be, as that is the only way to determine if the device is full or empty and if the flag status needs to be changed. Since the full flag needs to be referenced by the write side and the empty flag by the read side, each pointer must cross over to the other clock domain.

The whole point of the FIFO is to bridge a clock domain. Since now it is apparent that internal to the FIFO there need to be two data buses (the read pointer and the write pointer) crossing clock domains, this would seem to be a geometrically expanding recursive problem, but in reality it ends with these two crossings and the problem is manageable.

Assuming that the counters are at least two bits each, then the data coherence problem illustrated in Figure 7.20 will need to be solved. As an example of how the problem can occur, consider a four-bit pointer transitioning from seven to eight. In that case, there would be three one to zero transitions and one zero to one. If the faster one to zero transitions are all read with their new values but the zero to one transition is caught before reaching its transition threshold, the pointer would be latched in with a new value of zero instead of eight, resulting in an incorrect comparison with the other pointer.

In a binary counter, multiple bits can change on every cycle, leading to the problem of a clock domain crossing corrupting data. However, binary is not the only possible counter encoding. Gray counters, which have the distinguishing feature of only having a single bit change on any increment or decrement, can be used to cross the clock domain divide.

Gray codes for the values zero through 15 are shown in Table 7.2.

Table 7.2 Decimal, binary, and Gray values

Decimal	Binary	Gray	Decimal	Binary	Gray
0	0000	0000	8	1000	1100
1	0001	0001	9	1001	1101
2	0010	0011	10	1010	1111
3	0011	0010	11	1011	1110
4	0100	0110	12	1100	1010
5	0101	0111	13	1101	1011
6	0110	0101	14	1110	1001
7	0111	0100	15	1111	1000

Gray sequences do have a limitation: There are none for odd numbers. Only even numbers may be represented with a Gray sequence. With a binary counter, it is possible to wrap back to zero after any arbitrary value is reached. With a Gray counter, there is not any way to do that for odd numbers and retain the defining characteristic of a Gray sequence, which is that only one bit may change in any given cycle. This effectively limits FIFO depths to even numbers. If a memory cell is used rather than a register file, this is not a serious limitation, as memory cells typically not only always have an even number of locations, they are a power of two deep. A register file can be made any size but for use in a FIFO it must at least have an even number of elements.

Manually coing a Gray counter of arbitrary size is possible but tedious. A simple way of achieving the desired result is by using a Gray to binary encoder, as shown in Figure 7.24. That is a four-bit example, but it can easily be expanded to any size as long as the same structure, including the anomalous most significant bit, is maintained. Code for a binary to Gray encoder was shown in Chapter 5. The conversion could also be done in a function, as shown in Figure 7.25.

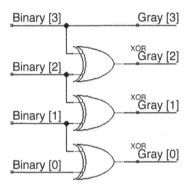

■ **FIGURE 7.24** Four-bit binary to Gray converter

```
parameter SIZE = 3;
reg [SIZE - 1 : 0] BIN, GRAY;

always_comb GRAY = B2G(BIN);

function [SIZE - 1 : 0] B2G(input [SIZE - 1 : 0] BIN);
  B2G[SIZE - 1] = BIN[SIZE - 1];
  for (int I = SIZE - 2; I >= 0; I--)
    B2G[I] = BIN[I] ^ BIN[I+1];
endfunction
```

■ **FIGURE 7.25** A function for binary to Gray conversion and a call to that function

A Gray counter could be created by putting a binary to Gray converter on the output of a binary counter and a Gray to binary converter on the input. A Gray to binary converter is shown schematically in Figure 7.26. The converter is a ripple process, so its

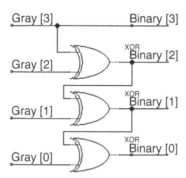

■ **FIGURE 7.26** Gray to binary converter

delays growth linearly with counter size, or logarithmically with FIFO depth.

As will be seen shortly, there is no need to use the Gray to binary converter, although its use can be convenient. If a Gray counter is wanted rather than just a Gray conversion of a binary counter and coding one of the appropriate size is too tedious, some synthesis tools will generate one automatically when instructed to do so and given a circuit description coded as a state machine. A function for a Gray to binary converter is shown in Figure 7.27. Although the function appears on both sides of an assignment operator, it is not recursive and is synthesizable.

```
function [SIZE - 1 : 0] G2B(input [SIZE - 1 : 0] GRAY);
  G2B[SIZE - 1] = GRAY[SIZE - 1];
  G2B[SIZE - 2] = GRAY[SIZE - 1] ^ GRAY[SIZE - 2];
  for (int I = SIZE - 3; I >= 0; I--)
     G2B[I] = G2B[I + 1] ^ GRAY[I];
endfunction
```

■ **FIGURE 7.27** Code for a Gray to binary conversion function

A Gray count can be sent across an asynchronous divide without risk of data coherence corruption, but replacing binary address pointers with Gray would create a new problem.

The issue can be illustrated by the following example using an eight-deep FIFO with four-bit Gray address pointers. There are eight writes followed by seven reads. With both pointers initialized to zero, this will leave the write pointer with a value of 1100 and the read pointer stopped at 0100. They are equal except for the MSB. According to the empty/full algorithm used with binary pointers, the FIFO is full, but it is not. It only has one word remaining in it.

Even worse, using the three least significant bits of the write pointer as the write address will cause the ninth word to be written to the eighth address rather than the first; thus, overwriting data and causing irrecoverable errors. This problem arises because Gray codes are symmetrical about their midpoint rather than repeating from the beginning as binary does. Comparing Gray pointers works for determining that the FIFO is empty but fails for determining if it is full. Further, using Gray codes as an address leads to data loss.

A solution to making the sequence repeat from the beginning rather than backtrack about the midpoint would be to replace the second most significant bit with its inverse when the counter is in the second half of its sequence. While this sounds complicated, in practice all it means is replacing that bit with the XOR of itself and the most significant bit.

The count values, Gray coding, and the desired memory addresses are shown in Table 7.3. The circuit for accomplishing

Table 7.3 Gray code and repeating address sequence					
Decimal	**Gray**	**Address**	**Decimal**	**Gray**	**Address**
0	0000	000	8	1100	000
1	0001	001	9	1101	001
2	0011	011	10	1111	011
3	0010	010	11	1110	010
4	0110	110	12	1010	110
5	0111	111	13	1011	111
6	0101	101	14	1001	101
7	0100	100	15	1000	100

■ FIGURE 7.28 Converting Gray sequence to repeating address pointer

this is shown in Figure 7.28. Once the Gray count has been made, the additional cost is only a single XOR gate.

With the second to most significant bit inverted half the time, the algorithm for determining if the FIFO is full needs another adjustment. When the FIFO is full, the most significant bits should be opposite. This is unchanged from a binary comparison. However, when the most significant bits are opposite, using the bit inversion scheme illustrated in Figure 7.28, the second to most significant bits of one of the pointers will have been inverted. Thus, when the FIFO is full, the two most significant bits will be opposite and rest the same.

This comparison algorithm works, but the modified sequence is no longer a Gray code. Attempting to use the N + 1 bits of a Gray counter with the second to MSB inverted to cross the clock domain boundary would lead to a high failure rate, as there would again be the potential for catching one bit with a new value and another with an old one when both are transitioning simultaneously.

However, there is no need to use the same bit pattern for addressing the memory and for comparing to the pointers to determine empty/full status. The two counters just need to run in lock step. Binary counters can be used for the addresses. Each counter then needs a Gray converter. Gray signals cross the clock domain boundary in each direction. Once synchronized with the other clock, the Gray code can be modified to change the second to the most significant bit as shown in Figure 7.28. Using plain binary for the address has the additional advantage of speed. Since the memory is likely to be the limiting factor on the entire FIFO design, any delay on converting the address will be in the critical path. Using plain binary for the address moves the conversion to Gray to be parallel with the memory access rather than in series with it.

A block diagram of an addressing algorithm and full flag logic is shown in Figure 7.29. The two discrete XOR gates would best be incorporated into the full flag comparator design but are shown on the diagram to clarify the intent. In the comparator, the second to MSB of the Gray buses would be replaced by the XOR gate outputs. For reliable operation, a two-stage synchronizer will be needed in the Write Clock Domain on the Gray bus coming from the Read Clock Domain. It is not shown in the diagram, but one is included in Figure 7.30 where the data flow in the opposite direction is illustrated. The memory cell is shown in both diagrams. There is only one memory cell in the design. It is not replicated for both paths.

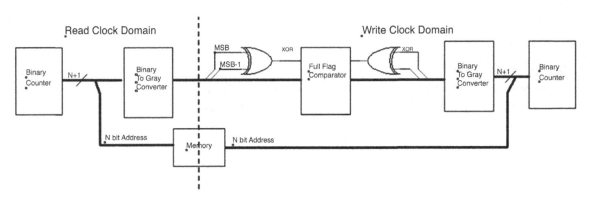

■ **FIGURE 7.29** Addressing and full flag logic; synchronizer not shown

An alternative algorithm is to convert the N + 1 bit Gray bus back to binary after crossing the clock domain divide. This would take a few more gates but it would not likely slow the circuit down because it would be done in parallel to the memory access. It has the advantage of making the flags easy to adjust. The FIFO can

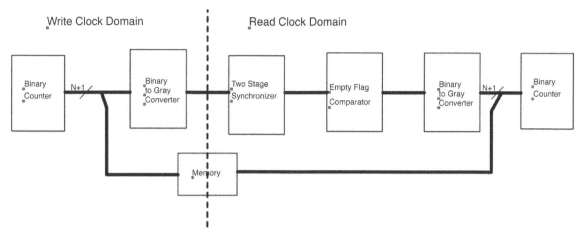

■ **FIGURE 7.30** Empty flag data flow

be designed to signal back to the reading and writing devices its status before it is totally full or empty. For example, the FIFO may be set to signal to the write side when only two words remain free. This feature can be essential in a pipelined operation when the data flow cannot respond instantaneously to a flag.

Because the empty flag is set when the pointers are equal for both binary and Gray codes, no extra processing is needed. If only an empty flag is needed and the receiving side will never need to know if the FIFO is almost empty, the fastest and easiest thing to do is to compare Gray codes. Otherwise, converting the Gray code from the write side back to binary may be the more practical approach.

FIFO operation and throughput

The FIFO described in this section is designed to bridge the gap between two circuits or systems operating asynchronously. Several different scenarios will be examined here to see if they can lead to failures when such a FIFO is used.

A common application of a FIFO is to buffer a burst of data for processing in a slower system. The transmitting clock can be many times as fast as the receiving clock. Part of the FIFO is a multibit two-stage synchronizer, exactly like the one of Figure 7.19 that will not work in the general case for preventing errors when changing clock domains. The difference is that the synchronizer in the FIFO only operates on Gray data rather than being required to operate on any random bit pattern.

Given that there can be a large difference between the two frequencies, the faster one can change several times before being read by the slower one. The two-stage synchronizer can be used on the parallel pointer because only one bit will be changing at any given time. But with multiple bits of the faster side having changed between reads of that bus by the slower side, will there again be a potential for loss of coherency?

The answer is unequivocally no. Since it is a Gray code that passes through the synchronizer, only one bit can be changing at any instant in time. No matter how many times other bits have changed in the past, at most one bit risks being caught by the receiving side clock. All the others will be steady state. Since they are not transitioning, they have no metastability risk. Loss of coherency is not a risk.

Passing through the synchronizer will add latency to the pointers. Does this add some risk of underflow or overflow, since each side may be using old, stale pointer values when determining if the FIFO is full or empty? Even with a synchronizer, can an error on the one bit that is changing cause a false reading that will lead to underflow or overflow? These questions will be answered by examining the operation of each flag under different conditions.

The empty flag is generated on the read side. It uses the synchronized, Gray-coded write pointer. Given that the write pointer may be operating much faster than the read pointer and it will go through additional delays in the synchronizer, the empty flag logic will sometimes be using old values of the write pointer.

The empty flag is set when the read pointer catches up to the write pointer. If the value of the write pointer used for the comparison is stale, the only possibility is that more writes have been done.

Using an old value to set the empty flag means that the empty flag may be asserted or remain asserted when the FIFO is no longer empty, but it can never result in the empty flag failing to be set when the FIFO really is empty.

It is failure to set the empty flag when needed that leads to underflow error. Keeping it set when it could be cleared can only lead to reduced throughput, never to data corruption due to reading from an empty FIFO.

The full flag is set when the write pointer catches up to the read pointer. It is generated on the write side, so it may be using stale read data. A stale read value can only mean that fewer reads are indicated than have actually been performed, so when the write pointer catches up to the old value of the read pointer, the full flag may be set prematurely but it can never fail to be set when it needs to be set. Once again, the only possible risk is loss of throughput, not loss of data.

Using possibly stale pointers is pessimistic, in that it can indicate that the FIFO is empty when data are available or full when space is available, but it does not add risk of the systematic failures that would happen if overflow or underflow were to occur.

Because only one bit at a time is changing in each synchronizer, there is no risk of loss of data coherency in the pointers, but synchronizers do not guarantee anything other than a low probability of a metastable output. What comes out of a synchronizer will be a legal logic value, but it could equally be the old value or the new one.

There are then four scenarios: on both the read and write side, either the pointer coming through the synchronizer will have an updated value or it will have the previous value. Because of the Gray code, those are the only possibilities. With only a single bit transitioning at a time, having a garbled, random value coming out of the synchronizer is never a possibility.

If the new values come through the synchronizers, then the flag logic will be using the latest data available, which is what is intended. If either or both synchronizers pass through the old value, the only penalty is an additional cycle of latency before the new value comes through on the next cycle. Following the same logic as just described for stale pointers, the worst that a synchronizer can do is make a pointer one cycle more stale, again decreasing throughput but nothing worse. The synchronizers can never output a value that would cause a FIFO to fail to set a flag when needed or clear a flag when still needed.

FIFO depth

Once the generic, scalable FIFO has been made, using it requires calculating how big it needs to be. Width is generally obvious: it needs to be as wide as the data paths. Depth requires some calculation.

In the long term, data into the FIFO must be equal to data out. In the short term, there can be vast differences between the data rates. A FIFO must be deep enough to buffer the worst-case short-term discrepancy between the two rates.

A typical application for such buffering would be in a communications system where many devices receive a burst of data from a single high-speed link, which they must process before the next burst comes along. The amount of data that needs to be buffered in the FIFO is the total data burst minus the amount that gets processed during the data burst. The number processed during the burst time is

$$N = B \times \frac{T1}{T2}$$

where B is the number of elements written during the burst, $T1$ the period of the write clock, and $T2$ the period of the read clock.

A numerical example of this would be a burst of 10 bits at 10 MHz, which is read by a 1 MHz system. In this system, a nine-bit deep FIFO would be needed, as only one bit would be processed during the burst and the other nine need to be buffered in the FIFO until the receiving side can get to them.

Because engineers work more often with frequencies than with periods, the formula for how deep a FIFO needs to be is commonly written as a function of frequencies

$$\text{Depth} = B - B \times \frac{f2}{f1}$$

where B is again the number of elements in a burst, $f2$ the output frequency, and $f1$ the input frequency.

Using the depth equation and some frequencies found in telecommunications systems, if the common bus is operating at 192 kHz and subscriber cards are operating at 8 kHz, each subscriber card would need to buffer $(B - B/24)$ elements. Assuming a burst is 192 bits, in the absence of any other complications the FIFO would need to be able to store 184 bits. Rounding up to the nearest power of two, a 256-bit deep memory or register file would most likely be the practical choice.

A common complication is clock jitter. Jitter is a short-term variance on the clock frequency. It may be intentionally introduced, as is done in spread-spectrum systems, or there may just be some required tolerance for jitter that occurs naturally with temperature variation and other circuit noise. An example of the latter is the PCI Express standard, which calls for components to tolerate up to 6000 PPM (0.6%) of jitter.

If the system must be able to tolerate 1% jitter on each side, in the worst case there will be a 2% difference between the transmit and receive sides. Assuming again the worst case, that the sender is operating 2% faster than the receiver, the buffer depth will need to be increased by that amount. This only works for short-term jitter that self-corrects. If there is any long-term frequency discrepancy, no amount of buffering will do. Overflow or underflow will be inevitable.

Since FIFOs typically are sized to the next-biggest power of two, it is rare for jitter to actually force an increase in FIFO depth.

A complete, hierarchical implementation of the FIFO is included in Chapter 12.

HIGH-SPEED ASYNCHRONOUS SERIAL LINKS

Previous sections in this chapter have provided design solutions for slow-speed single-bit asynchronous inputs and parallel clock domain crossings when there are asynchronous clocks on both sides.

A very common situation is to have a single-bit-serial data link with only the data being sent. No clock signal is sent along with the data. This arrangement is used in all long-distance links, whether long distance in terms of a few meters of cable connecting computer peripherals over a Universal Serial Bus or interplanetary radio communications. Some method of asynchronous serial communications is built into virtually every computer system or processor.

With low-speed data, simply avoiding metastability propagation via a two-stage synchronizer is good enough. If the new datum is not caught in one-clock cycle, it will be caught in the next. Such an approach will not work as data rates increase. While a FIFO could be made to work if the incoming serial data were accompanied by a clock, transmitting a clock along with the data stream is not viable either technically or economically. A circuit that will lock onto the data frequency is needed, and the design principles for creating such a circuit will be discussed in this section.

In order for such a data link to work, it must operate within some specified set of parameters. Without knowing even the nominal operating frequency, it would not be possible to lock onto the data stream. Both sides of the data link will need to stay within some percentage of the nominal frequency. There will also of necessity be some sort of a framing protocol.

Although the two sides of the link will be operating at the same nominal frequency, they will never operate at precisely the same frequency. Each will have its own time base, normally provided by a quartz crystal. As discussed earlier in this chapter, two crystals will never match to infinite precision. One will always be a little faster than the other. In the absence of a protocol for locking onto the transmitted frequency, inevitably either some bits would be read twice (if the receive clock were faster than the transmit) or some bits would be skipped (if the transmit clock were faster than the receive).

There are many protocols for serial data transfer. While differing in detail, they all have the common elements of a specified frequency tolerance and a framing protocol. An example of such a system will be developed here based on the DMX512 standard [5], which is a serial protocol maintained by the United States Institute for Theatre Technology. The choice of this particular standard is arbitrary. There are many others that would do as well. The full DMX512 specification contains numerous subtleties and electrical requirements that will not be examined in this example, which will deal with the logic of locking onto an asynchronous serial bus without getting bogged down in all the minutiae of any specification.

The DMX512 protocol calls for a 250 kHz bit rate. Both transmitters and receivers can have up to 2% variance from the center, meaning that there can be as much as 4% skew between them. In a DMX512 system, there is one transmitter. There is no feedback from receivers. Each receiver is programmed to extract data from one time slot. It is up to each receiver to lock on to whatever the transmitter sends.

A DMX512 frame starts with a "Space for Break," which is at least 23 bits of logic zero. It may continue indefinitely. Following that sequence, there are at least two bits of logic one. Then the data slot sequence starts. Each data slot has a logic zero start bit, then eight bits of data and at least two logic one-stop bits at the end. There may be many more stop bits between data slots. The first data slot is reserved for a start code that may be used to send signaling information to all the receivers. Following the first data slot there can be up to 512 more data slots. The number of data slots can change on the fly, all the way from zero to the full 512, as can the number of stop bits between slots, the length of the Space for Break and the Mark for Break. The structure of a DMX512 frame with two time slots is shown in Figure 7.31. Table 7.4 has a key for interpreting the components of the frame.

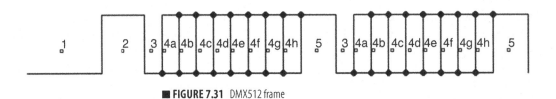

■ **FIGURE 7.31** DMX512 frame

Table 7.4 DMX512 frame components

Code	Meaning
1	Space for Break: Logic 0, at least 23 clock cycles
2	Mark for Break: Logic 1, at least two clock cycles
3	Start bit: Logic 0
4a–h	Data byte
5	Stop bits: Logic 1, at least two clock cycles

Because the transmit and receive clocks will not be operating at precisely the same frequency, a receiver detecting at least 22 logic zero bits in a row will interpret such a sequence as a valid Space for Break.

The framing sequence described here will always have a minimum number of logic zero to logic one and logic one to logic zero transitions. While the data packets would not necessarily have any transitions (all zeros and all ones being legal values for data packets), the combination of Space for Break, Mark for Break, Start bits, and Stop bits is enough for the receivers to lock onto the transmitter's frequency.

If all time slots are used along with the required minimum number of stop, space, and mark bits, the entire frame will be 5668 bits. Under worst-case skew conditions, a transmitter and receiver pair would have a discrepancy of over 226 bits by the end of the frame. Without locking onto the transmitter's frequency, extracting data packets would be impossible.

The general solution to this class of problem is to have a fast clock in each receiver. A counter is incremented on each rising edge of this clock. When transitions on the incoming data stream are detected, as a result of either data transitions or the framing sequence, the count is zeroed, thus, adjusting itself to the transmitter's frequency.

The Nyquist theorem would seem to allow this fast clock to be as little as twice the bit rate, but such a low multiplier would only work if the transmit clock could never be faster than the receive clock. This is not the case, as the DMX512 specification allows it to be up to 4% faster. In practice, almost all such locking circuits use a multiplier of 16, although slower and faster internal clocks could work. Figure 7.32 uses a factor of 16 for the fast clock.

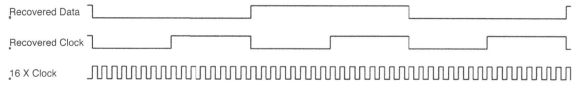

■ **FIGURE 7.32** Fast internal clock is used to lock on to incoming frequency

If the transmit and receive clocks were perfectly aligned, the receiver's counter would always roll over from 15 back to zero precisely when a new data bit arrives, as illustrated in Figure 7.33.

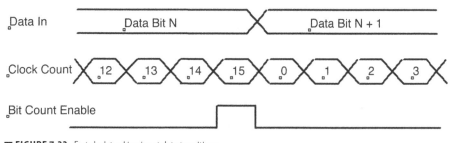

■ **FIGURE 7.33** Fast clock tracking input data transitions

In order to keep them aligned when the two frequencies do not exactly match, the count is set to zero when a data transition is detected. If the frequency differential is at its maximum of 4% and data constantly alternate between zero and one, it would result in the count being adjusted by one about every other data bit. If the transmit clock were faster, the count would jump from 14 to zero, and if the receive clock were faster, there would be a double zero each time an adjustment was made.

The received data, however, will not always alternate between zero and one. Within the data packets, there may not be any transitions at all. Although the data may stay constant, each data slot will still have at least one transition: If data are all zeros, there will be the zero to one transition of the stop bits; and if data are all ones, there will still be the logic zero start bit before the data packet. Thus, there is a limit to how long there can be between transitions that are used to adjust the count.

It is that limit that determines how much frequency tolerance the system can support. Data of all zeros following the logic zero start bit gives nine bits in a row of zeros. With a frequency difference of 4%, by the time the stop bit arrives the internal count will be off by 36% of a bit.

The calculation is similar for a data packet consisting of nothing but logic ones, but that condition can persist indefinitely as there may be many stop bits. However, there is no need to track which bit is being received when the packet is padded with additional stop bits. The data slot will always end eventually with either the logic zero of the next slot's start bit or the logic zero of a Space for Break. Whenever the next logic zero arrives, the counter can be reset. With the data packet set to all ones, it only matters that the internal count track the entire data packet of 8 bits plus two stop bits, for a total of 10 bits in a row of logic one.

After 10 bits with maximal frequency error, the two sides would be off by 40% of a bit. That too is a recoverable error. Anything less than 50% is recoverable.

Fifty percent is the limit because that is where the clock count adjustment algorithm would fail. This algorithm says to advance the bit count if a new data bit arrives when the clock count has exceeded its midpoint, but the clock count should return to zero without advancing the bit count if the clock count has not reached its midpoint. If the skew between the clocks builds up to more than 50% between data input transitions, the adjustment will come out backwards and the position in the frame will be lost. By keeping the maximum skew that can build up between transitions to 50% of a bit, adjustment to the count will always be in the right direction when the bit counter "springs forward" when the count is more than 50% and "falls back" when it is less. The algorithm is illustrated in Figures 7.34 and 7.35. In Figure 7.34, the transmitting clock is faster than the receiving clock. The internal fast count has only reached 13 when a new bit arrives. The response is to jump ahead to zero, incrementing the bit counter.

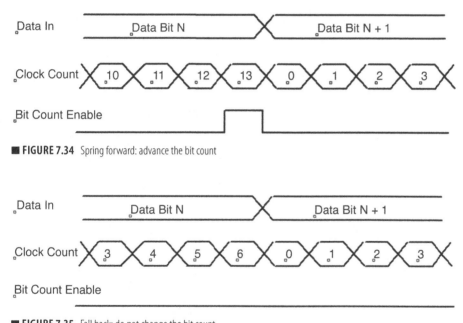

■ **FIGURE 7.34** Spring forward: advance the bit count

■ **FIGURE 7.35** Fall back: do not change the bit count

In Figure 7.35, the opposite situation is illustrated. The faster receiving clock has wrapped around and counted back up to six before a new data bit comes in. Because six is still less than halfway through the count, it falls back to zero when a new data bit is detected and does not send an enable to the bit counter.

Keeping the frequency tolerance to 2% ensures that the spring forward/fall back algorithm will always work. If the frequency tolerance were 3%, or a total of 6% difference between transmitter and receiver, after 10 bits in a row without a transition the skew could be up to 60% of a bit. Recovery then would be impossible: there would be no way of knowing if the bit counter should be incremented or not.

Because skew can also build up during the framing sequence, the standard requires that at least 23 bits of logic zero be sent, but states that receiving 22 is adequate to determine that a frame has started. This allows the slowest legal receiving clock to still detect the shortest legal framing sequence.

A block diagram of an asynchronous serial bus interface is shown in Figure 7.36. A complete implementation of a DMX512 receiver is included in Chapter 12. The DMX standard is somewhat more complicated that many serial protocols, since the number of time slots can change in every frame, as can the length of Space for Break, Mark for Break, and the number of stop bits.

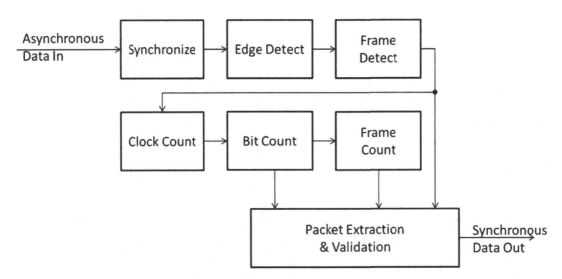

■ FIGURE 7.36 Asynchronous serial bus interface

■ SUMMARY

At least one of the techniques discussed in this chapter should be adaptable to any asynchronous interface problem.

Keeping designs totally synchronous will result in more reliable and robust products than the alternatives, but pressures to maximize performance and minimize power consumption sometimes lead designers to compromise on this principle. By at least keeping blocks locally synchronous and using the techniques described here when passing signals from one clock domain to another, circuit failures can be minimized. Chapter 9 includes more information on reducing power consumption while keeping circuits as synchronous as practical.

REFERENCES

[1] Cummings, C., Mills, D. "Synchronous Resets? Asynchronous Resets? I am so confused! How will I ever know which to use?" SNUG2002, San Jose, CA; 2002.

[2] Wellheuser, C. "Metastability Performance of Clocked FIFOs." Texas Instruments white paper scza004a; 1996.

[3] "CLOCK DOMAIN CROSSING CLOSING THE LOOP ON CLOCK DOMAIN FUNCTIONAL IMPLEMENTATION PROBLEMS." Cadence Design Systems, Inc. white paper; 2004.

[4] Cummings, C. "Simulation and Synthesis Techniques for Asynchronous FIFO Design." SNUG2002, San Jose, CA; 2002.

[5] American National Standard E1.11-2004, USITT DMX512-A, Asynchronous Serial Digital Data Transmission Standard for Controlling Lighting Equipment and Accessories.

Simulation, timing, and race conditions

ASIC design methodology using a hardware design language is dependent on digital simulation. This simulation is an imperfect representation of how a real circuit will perform, but it is the most practical technique for initial design verification.

Over the years numerous simulation algorithms have been developed. The thrust has always been to increase speed. The first digital simulators were time driven, where each node was evaluated at each instant of time. Since at any given time, only a small percentage of nodes will have changed from the previous instant, most of the computing effort was wasted. More modern algorithms are vastly more computationally efficient and modern simulators are capable of taking advantage of multiprocessor platforms, but all simulators share some fundamental characteristics.

Most fundamental of all is the concept of simulation time monotonically advancing from zero.

SIMULATION QUEUES

To advance monotonically from zero, simulators have a series of queues. All events that would occur simultaneously in the circuit under simulation are scheduled by being pushed into a queue.

The simulator will sequentially evaluate each, assigning the computed value to each node and working through the queue until it is empty. It will then be able to move on to the next time step, where the whole procedure will begin anew. A sequence of five queues with six events in each is shown in Figure 8.1. The host computer running a simulation will take real time to execute all the events in each queue, but simulation time will only advance each time a queue has been exhausted.

The order in which events are pushed into queues is indeterminate. While it will be deterministic in that the order of execution will be repeated if the simulation is run multiple times without any changes to the source code, the user cannot directly specify any preferred scheduling order.

Queues are dynamic in that events of one queue can cause new events to be scheduled in later queues. Once a time slot has been exhausted, however, it will never be revisited. Simulation time can only advance.

The diagram in Figure 8.1 is vastly simplified from what goes on inside a simulator. A SystemVerilog simulator has over a dozen queues for each time slot, each dedicated to different categories of events.

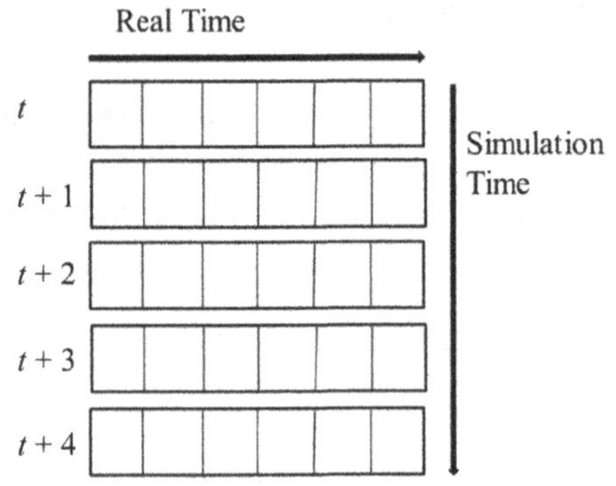

■ **FIGURE 8.1** Simulation queues: simulation time only advances after a queue has been emptied

The conceptual diagram of Figure 8.1 shows all queues being the same length. There is no requirement that they be so and having such a symmetrical sequence of events in reality would be extraordinarily rare.

Figure 8.1 illustrates a fundamental difference between simulation of a circuit description and physical circuits. The events in a simulation queue run sequentially. In a physical circuit, components can change state simultaneously.

RACE CONDITIONS

The implications of sequential simulation of parallel events can be seen with the RS latch models introduced in Chapter 7. For the NAND latch shown in Figure 8.2, setting R and S simultaneously from one to zero is entirely deterministic in simulation. First one of the gates will be evaluated, then the other. That a silicon device will have a race condition is not a factor for the simulator. Simulation will always show Q and Qb set to logic one. This does not match reality, where the actual outputs could take any value, including illegal, intermediate states.

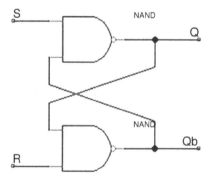

■ **FIGURE 8.2** NAND RS latch

Race conditions are not limited to instances of primitives. The code of Figure 8.3 also has a race. The outputs will both be the same, but whether they end up at logic one or logic zero will be determined by which signal propagates quicker, which will primarily be determined by circuit layout. Circuit behavior cannot be determined through behavioral code simulation for this class of problem.

```
`timescale 1 ns / 1 ns
module race (OUT1, OUT2, CLK, RST);
output OUT1, OUT2;
input CLK, RST;
reg OUT1, OUT2;

always @(posedge CLK or negedge RST)
  if (!RST) OUT1 = 0;
  else OUT1 = OUT2;

always @(posedge CLK or negedge RST)
  if (!RST) OUT2 = 1;
  else OUT2 = OUT1;

endmodule
```

■ **FIGURE 8.3** Race condition in behavioral code

Simulation of the code in Figure 8.3 will indicate that both out-
puts will always be logic one. However, simulation of the code of
Figure 8.4 will show that both outputs will always be logic zero.
Synthesis of the two modules will produce identical netlists.

```
`timescale 1 ns / 1 ns
module race (OUT1, OUT2, CLK, RST);
output OUT1, OUT2;
input CLK, RST;
reg OUT1, OUT2;

always @(posedge CLK or negedge RST)
  if (!RST) OUT2 = 1;
  else OUT2 = OUT1;

always @(posedge CLK or negedge RST)
  if (!RST) OUT1 = 0;
  else OUT1 = OUT2;

endmodule
```

■ **FIGURE 8.4** Same race, different outcome

Simulation races are not limited to feedback loops as shown in Figures 8.2–8.4. Another race is shown in Figure 8.5. In that example too, the simulation results can be changed by reordering the functional blocks. As written, the display message will execute. If the second initial block is made the first, the display statement will not run. The issue is the order in which events are pushed onto the queue for simulation time 10. With sequential blocks in a single file, one can be reasonably confident that the events will be scheduled in the same order in which they appear in the source code, although there is no guarantee that this will always be so. When multiple files are involved, the order of execution becomes less obvious.

```
module race2;
  reg a2;

  initial begin
    a2 = 1'b0;
    #10 a2 = 1'b1;
  end
  initial
    #10 if (a2) $display("This may not print");

endmodule
```

■ **FIGURE 8.5** Results will depend on queuing order

In the example of Figure 8.5, changing the display task to a strobe will not help, despite the fact that strobe statements go in a different queue, one that runs after the register assignments, although still at the same simulation time [1–3]. If the evaluation of a2 comes before a2 is assigned a value of logic one, the message will not be printed regardless of which task is used. Neither a display task nor a strobe task will be scheduled when the order of the blocks is reversed, although both would be scheduled with the code shown in Figure 8.5.

The evaluation of a2 can be forced into a different queue by adding an infinitesimally small delay, even a zero delay, as shown in

```
module race2;
  reg a2;

  initial
    #10 #0 if (a2)
    $display("This one is different. It will print");

  initial
    #10 if (a2) $strobe("This may not print");

  initial begin
    a2 = 1'b0;
    #10 a2 = 1'b1;
  end

endmodule
```

■ **FIGURE 8.6** Addition of a zero delay will put the evaluation into a different queue

Figure 8.6. Although the evaluation and subsequent display statement of the first initial block will still be at simulation time 10, the first block will not be scheduled until after both the second and third blocks are executed. "This one is different. It will print" will be displayed no matter what the order of the blocks is in the source code. Adding a zero delay moves it to a separate "inactive" queue that is evaluated only after all events in the "active" queue are assigned. It is still at simulation time 10, but not the same 10 as the other assignments.

DERIVED CLOCKS AND DELTA TIME

Figure 8.7 illustrates another scheduling issue, the use of a derived signal as a clock. The module has a master clock input, but it creates internally a logic signal of one half the master clock frequency and uses that signal as a clock. Two blocks are clocked with the master clock and one is clocked with the derived logic signal.

In simulation, it would appear that CLK and CLK2 are perfectly aligned. However, not only would there be some skew between them if this circuit was to be built, but there is also a difference in how signals from the different blocks are scheduled.

```
module race3(input CLK, RST, output logic [3:0] CNT,
  output logic MATCH);
  logic CLK2;
  const int TC = 15;

  /*When this block is last, MATCH is asserted when CNT = 15.
  When it is first, MATCH is asserted when CNT = 0.
  If all sequential assignments are made with non-blocking
  assignment operators, the order of the blocks does not
  matter and MATCH is always asserted when CNT is 15. That
  is also the case when all blocks use the master clock
  rather than having one use CLK2.*/
  always_ff @(posedge CLK, negedge RST) begin
    //Derived clock, 1/2 frequency of master clock
    if (!RST) CLK2 = 1'b0;
    else CLK2 = ~CLK2;
  end

  //This block uses the derived clock
  always_ff @(posedge CLK2, negedge RST) begin
    if (!RST)
      CNT = '0;
    else
      if (CNT == TC) CNT = '0;
      else CNT = CNT + 1;
  end

  //This block uses the master clock
  always_ff @ (posedge CLK) begin
  //Match signal is "and"ed with CLK2 so it
  //will be active for only one master clock cycle
    if (CNT == TC && CLK2) MATCH = 1'b1;
    else MATCH = 1'b0;
  end

endmodule
```

■ **FIGURE 8.7** Using a logic signal as a clock

Without any gate delays in the behavioral circuit description, there is nothing in the code to indicate when signals triggered from a clock signal should change. However, it would be incorrect behavior to have a signal triggered by a derived signal change before the original clock signal, even if all are nominally at the same simulation time.

In the code of Figure 8.7, it would be illogical and incorrect for CNT, which is a function of CLK2, which is itself a function of CLK, to change state before CLK. To avoid that sort of illogical evaluation order, CLK2 is a "delta cycle" later than CLK. A delta cycle is a vanishingly small time period that puts events into a new queue without advancing simulation time.

CNT is also updated a delta cycle after CLK. This means that there is a race between updating CLK2 and CNT. Which gets updated first is a matter of which gets queued first. In simulation, this can be controlled by changing the order in which blocks appear, if they are all in the same module. If they are scattered across a large design hierarchy and subject to different enabling events, controlling or predicting behavior will be harder.

In a physical circuit, behavior would be layout-dependent. MATCH might be asserted when CNT is 15. It might be asserted when CNT is zero. The circuit might not work at all due to setup or hold violations.

While delta cycles are a software concept, physical circuits also have delays that can cause havoc when designs are done without keeping delays in mind. Using a logic signal such as CLK2 in Figure 8.7 as a clock will create a second clock domain in the module. It will not be possible to reliably read signals created by one clock with the other.

In Figure 8.7, seemingly correct operation can be obtained in simulation by changing all the assignment operators to nonblocking ones. With nonblocking assignments, evaluation of the right-hand side of equations is done in one queue with the results assigned in a later nonblocking assignment queue, so there is never any mixing of old and new values in the assignments. While use of nonblocking assignment operators for sequential blocks is recommended, doing only that would still leave the circuit description with two incompatible clock domains.

Instead of using CLK2 as a clock signal, reliable operation can be obtained by using it as an enable. The circuit description is reworked in Figure 8.8 to be fully synchronous and to use the recommended assignment operators. To have the MATCH signal active for the second half of CNT = 15, the phase of the signal used to validate MATCH was also changed.

The synchronous design of Figure 8.8 accomplishes what the design of Figure 8.7 intended to do without race hazards. Simulation

```
module race4(input CLK, RST, output logic [3:0] CNT,
  output logic MATCH);
  logic CLK2;
  const int TC = 15;

  //Enable every other clock cycle
  always_ff @(posedge CLK, negedge RST) begin
    if (!RST) CLK2 <= 1'b0;
    else CLK2 <= ~CLK2;
  end

  always_ff @(posedge CLK, negedge RST) begin
    if (!RST)
      CNT = '0;
    else
      if (CLK2 && CNT == TC) CNT <= '0;
      else if (CLK2) CNT <= CNT + 1;
      else CNT <= CNT;
  end

  always_ff @ (posedge CLK) begin
  //Match signal is "and"ed with CLK2 so it
  //will be active for only one master clock cycle
    if (CNT == TC && !CLK2) MATCH <= 1'b1;
    else MATCH <= 1'b0;
  end

endmodule
```

■ **FIGURE 8.8** Synchronous design with nonblocking assignment operators

■ **FIGURE 8.9** Simulation of the final synchronous design

of the design is shown in Figure 8.9. While the same output can be obtained with the code of Figure 8.7, not only will the code of Figure 8.8 simulate as shown, but a circuit synthesized from it will perform identically.

ASSERTIONS

Assertions are a SystemVerilog construct added to the language to simplify verification code. They are not synthesizable and are most commonly used in verification programs. They can, however, be embedded in design code, where they will trigger events during verification but be ignored by synthesizers.

There are two fundamental types of assertions, immediate and concurrent. Immediate assertions, such as standard Verilog assignments, are triggered when they are encountered in the code. Concurrent assertions operate over multiple clock cycles. That is, they run concurrently with the design being verified.

An application for immediate assertions could be to determine if the flags on a FIFO are operating properly. Using the FIFO described in Chapter 7 and implemented in Chapter 12 as an example, the following assertion would display an error message when the EMPTY flag was not set but the pointers indicate that it should be.

```
always_comb assert (RD_GRAY == WR_GRAY && !EMPTY)
$display("Error: at time %d empty flag should be set but is not", $time);
```

Like blocks, assertions can be named. In the lines below, the assertion that checks the EMPTY flag is given the name CHECK_FLAG.

```
always_comb CHECK_FLAG: assert (RD_GRAY == WR_GRAY && !EMPTY)
$display("Error: at time %d empty flag should be set but is not", $time);
```

An application for a concurrent assertion could be to check a handshake protocol between two modules. For example, one module could set a Data Valid signal high. The other module would be waiting for that to occur. When it does, it would set its Acknowledge line high. The first module would then respond to that by clearing Data Valid, after which the Acknowledge line should be cleared. Concurrent assertions could be used to verify that the protocol is completed in a given number of clock cycles.

In concurrent assertions, delay times are given in numbers of clock cycles, indicated by ##. In the following assertion code, after DATA_VALID goes high, ACK should go high two cycles later. Then DATA_VALID should be removed after no less than one cycle and no more than five. Finally, ACK should be cleared after three more cycles.

Concurrent assertions work with "properties." The property specifies the sequence of events for a concurrent assertion. Both assert and property are SystemVerilog keywords.

```
always @(posedge CLK)
  assert property (DATA_VALID ##2 ACK ##[1:5] !DATA_VALID ##3 !ACK)
  else $display("Error: handshake failed at time %d", $time);
```

■ SUMMARY

Simulation does not always accurately represent circuit behavior. The discrepancies are particularly insidious with race conditions, where simulation will often indicate a stable, predictable outcome that does not reflect reality.

Such discrepancies can be minimized through use of synchronous design techniques. Avoiding any use of logic signals as clocks can eliminate most such issues. Using the proper assignment operator can eliminate most of the rest.

A fully synchronous design that always uses nonblocking assignment operators for sequential blocks and blocking assignment operators for combinational ones will not normally experience simulation/synthesis mismatches, but combinational feedback such as is used in the RS latch of Figure 8.1 will still have a potential for such problems. For this reason, many fabricators forbid combinational feedback loops outside of instantiated components such as flipflops. Even the use of latches is often discouraged, despite their inclusion in many component libraries.

Assertions and their associated properties are used to simplify verification code. They may be embedded in circuit description source code, where they can print error messages or trip an abort to the simulation when the design is not working as intended. They are ignored during synthesis and have no impact on the circuit produced.

REFERENCES

[1] Cummings, C and Salz, A. "SystemVerilog Event Regions, Race Avoidance & Guidelines." SNUG Boston, 2006. http://www.sunburst-design.com/papers/Cummings SNUG2006Boston_SystemVerilog_Events.pdf

[2] Mueller, W, et al. "The Formal Simulation Semantics of SystemVerilog." Proceedings of the Forum on Specification & Design Languages, 2004. http://adt.cs.upb.de/wolfgang/fdl04.pdf

[3] IEEE 1800-2012 SystemVerilog Language Reference Manual.

Chapter 9

Architectural choices

Many implementations of a design concept may be logically correct, but some will be better than others when more factors are taken into account. Power consumption, area, speed, reliability, and time to market are some of the considerations that can influence architectural choices.

FPGA VERSUS ASIC

Verilog and SystemVerilog are equally useful for ASIC and FPGA designs, but not all design techniques will apply equally to both. FPGA designs are typically more constrained than ASICs. Examples of this are that some FPGA technologies do not support internal Tri-state buses, some do not have latches in their libraries of components, and combinational feedback may be totally prohibited. The designer will need to be cognizant of the limitations of the targeted technology.

Because clock routing is built into FPGAs before the design begins, an FPGA will have a hard limit on the number of clock domains. While this number may be large, it cannot be exceeded. Custom circuits have no such limitation. Again, the FPGA designer must know the limit to avoid creating a design that cannot be implemented. Because the built-in clock networks are designed to have low skew, deliberately delaying a branch of the clock to allow a difficult path to have more time is not an option in an FPGA, whereas the ASIC designer can sometimes do this in layout to allow a design to meet timing.

Coding examples throughout this book have assumed that an asynchronous reset is the proper method of putting registers into an initial, known state. This does not necessarily apply to FPGAs, as FPGA registers may be configured into their initial state at power up. Furthermore, some FPGA constructs are optimized to work without resets (Xilinx® shift registers are a well-known example of this) and adding a reset will prevent the optimized configurations from being used, consuming far more hardware resources than necessary.

ASIC libraries typically support both asynchronous preset and reset inputs on flipflops. The top-level design can have as many asynchronous preset and reset inputs as is convenient. In an FPGA, the number of such inputs may be severely limited and an individual register may be limited to one such input. In some FPGA technologies, a register may have an asynchronous preset or a reset, but not both.

FPGA sequential blocks are often written with just a clock signal with other controls done synchronously. This is illustrated in Figures 9.1 and 9.2, where the ASIC style is written first, followed

```
//ASIC style: asynchronous active low reset and preset inputs
//EN is a synchronous logic signal
module ASIC_P_R(input CLK, RST, PRE, EN, D, output logic Q);
  always_ff @(posedge CLK, negedge RST, negedge PRE)
    if (!RST) Q <= 1'b0;
    else if (!PRE) Q <= 1'b1;
    else if (EN) Q <= D;
    else Q <= Q;
endmodule
```

■ **FIGURE 9.1** ASIC coding style for a D flipflop with asynchronous preset and reset inputs

```
//FPGA style: for optimal resource usage, always code
//registers so that reset overrides preset and both have
//priority over clock enable (CE).

//Preset and reset are active HIGH in this example. ASIC
//style almost always has asynchronous preset and reset
//active low.

//No final "else" clause needed. If the clock is not
//enabled, no action will occur.
module FPGA_P_R(input CLK, RST, PRE, CE, D, output logic Q);
  always_ff @(posedge CLK)
    if (RST) Q <= 1'b0;
    else if (PRE) Q <= 1'b1;
    else if (CE) Q <= D;
endmodule
```

■ **FIGURE 9.2** FPGA style flipflop

by the FPGA style. An FPGA that does not follow this coding template may be highly inefficient, as it will use logic resources to implement reset and preset functions that would otherwise be done with a dedicated preset and reset. The FPGA example also takes advantage of the built-in clock enable (CE) signal that is present in some FPGA technologies. Rather than use a multiplexor to feed back the output when no output change is desired, with this type of technology, if the clock is not enabled, the register will hold state.

FPGA designers may be tempted to instantiate cells that are specific to the target technology, but an ASIC designer will tend to write portable code that can be reused when the design migrates to a newer semiconductor process. Since FPGA designs also sometimes migrate to higher-performance ASIC technologies, limiting the use of instances of vendor-specific cells may be more productive in the long run. However, resisting that temptation is not always the best procedure, as the FPGA-specific cells may be higher performance than anything the designer will come up with and will allow the design to be completed sooner.

An FPGA design may need more pipelining stages than an ASIC. This is because FPGAs typically have their random logic functions configured into blocks with four or five inputs, which are associated with a flipflop. While an ASIC can have as many components as are needed in a combinational path, going from one FPGA logic block to another is a slower process. Accordingly, using the flipflops associated with each logic block may allow the FPGA to run at the required clock speed, but at the cost of more latency.

Clock-gating techniques for power reduction described later in this chapter may be unavailable in FPGAs. Even if they are possible, they tend to be less effective in FPGAs than in custom-mask devices, as the clock routing in an FPGA is fixed when the die is created and the static to dynamic power ration tends to be higher in an FPGA than in a custom-mask device.

DESIGN REUSE

Time to market is intertwined with design reuse. Reusing an existing design or piece of a design, even if it is not a perfect fit, can be a rational choice if it can get a working product onto store shelves quicker and reduce development costs. A corollary is that partitioning designs into scalable, reusable subdesigns with standard interfaces may incur additional costs the first time the strategy is implemented but will result in savings in subsequent iterations.

Design reuse can range from something so simple as instantiating two existing four-bit counters instead of writing a model for an eight-bit counter to buying and instantiating an entire processor core. In the first example, the savings would be trivial, as writing an eight-bit counter in Verilog is trivial for an experienced user, but creating an application-specific processor can be a major undertaking.

One of the keys to facilitating design reuse is to partition subdesigns into blocks that have a high potential for reuse in other systems.

PARTITIONING

Keeping each circuit description file down to about two pages makes the files easy to maintain and understand, but in a design with tens of millions of gates, that makes a lot of files. Guidelines for organizing designs and files are presented here.

Standard design methodology calls for registering module outputs and keeping combinational logic in the same module as their destination registers. Operational necessity will sometimes cause this guideline to be violated, but such necessity is rare. Following it will almost always allow synthesizers to produce optimal results.

An example of poor partitioning is shown in Figure 9.3 [1]. If the modules are synthesized individually, optimization across the entire logic path will not be possible. The result will all but inevitably be a circuit that is both big and slow. All the irregularly shaped blocks in the following diagrams represent combinational logic gates.

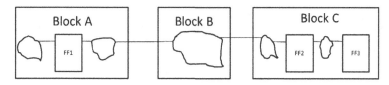

■ **FIGURE 9.3**   Poor partitioning with combinational paths split between modules

The partitioning shown in Figure 9.3 can be improved by moving combinational logic from Blocks B and C into Block A, eliminating Block B entirely. This keeps each combinational path in a single file, where the entire path may be optimized. Figure 9.4 shows this better approach to partitioning.

While it does keep all associated combinational logic together, it still fails to register block outputs, which is usually recommended practice. Doing so provides a standard interface, allowing the designer of each block to depend on having the full cycle for arriving signals rather than having just some unknown portion of a cycle.

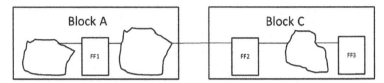

■ **FIGURE 9.4** Better partitioning: no combinational paths split between blocks

If Block B contains a lot of complex logic that would make the other files too long and unwieldy if merged, offloading the combinational circuit descriptions to tasks or functions that can be called from Blocks A or C would be a practical approach. Task and function subroutines are covered in Chapter 6.

Figure 9.5 shows the recommended method: keep combinational logic with associated registers and register all outputs.

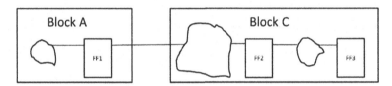

■ **FIGURE 9.5** Recommended design partitioning: all block outputs registered, no split logic

Another recommended methodology is to avoid any "glue logic" at the top level. Having only instantiated blocks there often means that the top level does not need to be synthesized at all, and if it does, it will just be to insert buffers as necessary to meet electrical requirements. However, rigorously following this guideline can make it impossible to reuse existing blocks.

An example is shown in Figure 9.6, where a single gate is used at the top level to connect together three blocks. It is generally desirable to move any such combinational logic into the receiving block, as shown in Figure 9.7. That would require changing both the internal code of Block C as well as its ports, which is not always possible. Writing a separate wrapper module for Block C in which the original design is instantiated along with the code for the logic from the top level is a way of finessing this problem, although at the cost of some design elegance.

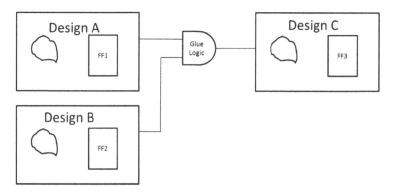

■ **FIGURE 9.6** Top-level design with glue logic

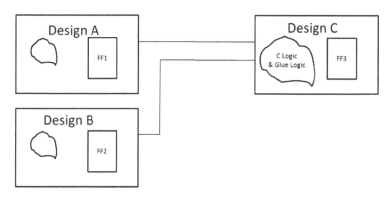

■ **FIGURE 9.7** Top-level design as a purely structural netlist

When partitioning a design, it is also wise to anticipate where difficult timing paths are likely to be and isolate them into as small as possible modules. Those modules can then be synthesized for optimal speed, which will cause their area to expand. The rest of the design can then be synthesized for optimal area.

When care is not taken in doing the initial design, particularly if a bottom-up approach is used, ending up with a system like the one shown in Figure 9.8 is possible [2]. With a chaotic communications structure such as is shown there, design maintenance becomes difficult. When the design gets to layout, performance too will suffer, as routing will necessarily require long and convoluted interconnects. A change in one module may necessitate changes throughout the entire structure, instead of being localized to one block and its nearest neighbor. The implications of

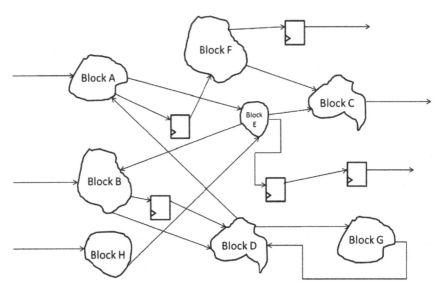

■ FIGURE 9.8 Chaotic communications lead to difficult design maintenance and layout

changing one subdesign will not be obvious and the full impact of a change may not be found for a long time.

The partitioning of Figure 9.8 also violates all the guidelines offered in Chapter 8. Some outputs are registered, some are not. Combinational paths are split between design units. There is a combinational feedback. While no one would set out to produce such a design, failure to follow established top-down methodology with consistent intermodule interfaces does end up with a structure looking altogether too much like that shown.

Instead of such an incoherent structure, a more idealized flow is shown in Figure 9.9. While not all designs will be amenable to such a simple and straightforward flow, making an effort at the start of a project to partition it for simple data interchange will pay off before the design is finished.

To facilitate creation of such a regular structure, the number of any design units at any given level of hierarchy is best limited to a modest number. Psychology experiments dating from the 1950s have shown that seven discrete objects, plus or minus two, is what the human brain can keep track of at a time [3]. This leads to the

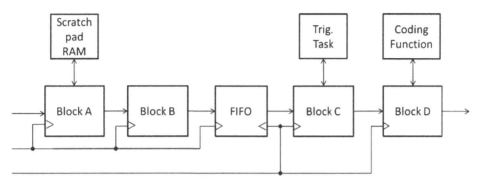

■ **FIGURE 9.9** Design partitioned for optimal connectivity and ease of maintenance

"rule of seven," that the optimal number of instances at any level of hierarchy is seven, plus or minus two.

Small as it is, even the FIFO design developed in Chapter 7 and implemented in Chapter 12 has enough functional blocks that creating one flat design would make it awkward and hard to understand. Figure 9.10 shows all the functional blocks of the FIFO in one flat array.

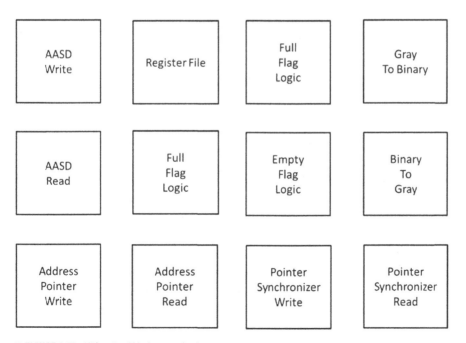

■ **FIGURE 9.10** All functional blocks at one level

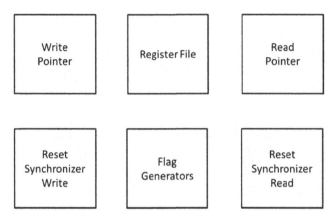

■ **FIGURE 9.11** Following the Rule of Seven makes the design easier to visualize

In Figure 9.11, the FIFO hierarchy is reworked to follow the rule of seven. This is the structure that is implemented in Chapter 12. There are six top-level blocks, with the others instantiated in one of these six. Such a structure is easier to visualize and maintain. It also provides reusable design modules that may prove useful in other projects.

Following the Rule of Seven at all levels of hierarchy is another recommended design technique.

AREA AND SPEED OPTIMIZATION

Engineers have long known that speed can be traded for area. With different optimization constraints set, synthesizers can produce a myriad of different implementations from one set of Verilog files. In general, the tradeoffs involve adding redundancies and stronger drivers to get greater speed at the cost of increased area. The classic if somewhat idealized "banana curve" is shown in Figure 9.12. In reality, such a smooth line is highly improbable and there may be local minima, but getting a 2:1 improvement in either speed or area is within the realm of probability. Not all designs will do that well. The greater the ratio of registers or memory to combinational logic, the less the synthesizers will be able to move along the banana curve, changing the implementation to gain either area or speed.

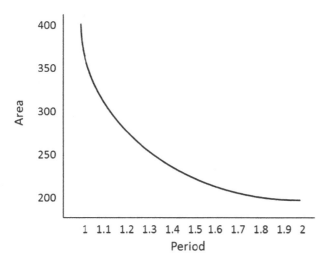

■ **FIGURE 9.12** Trading area for speed

That much optimization can be achieved without any change to the circuit descriptions. Coding changes can provide far more optimization. An example of how a small change in a circuit description can produce a significant change in the implementation is shown in Figure 9.13 [4]. In the two modules shown, the first will cause eight 16-bit adders to be generated. By removing the addition operation from the loop, the second will only make one. Minimizing code in a loop can produce outsized gains in area reduction.

The code in Figure 9.13 was simplified from the original design of [4] to emphasize the benefit of reducing logic in loops. The design is part of a computer interrupt controller. If implemented as shown in Figure 9.13, when there are multiple concurrent interrupts, the one with the highest numerical value will be the one selected. This may not suit all designs.

In both implementations, latch generation is avoided by assigning a variable a default value before the loop is encountered. This coding style may not be accepted by all synthesis tools. Removing the default assignment and adding an "else" clause following the "if" inside the loop will always work.

```
module IRQ_ADDR1(input [7:0] IRQ, input [15:0] BASE_ADDR,
  output logic [15:0] ADDR);
  const logic [15:0] OFFSET [0 : 7] = {4, 8, 12, 16, 20, 24, 28, 32};
  always_comb begin
    ADDR = BASE_ADDR;
    for (int I = 0; I < 8; I++)
      //Eight adders will be made
      if (IRQ[I]) ADDR = BASE_ADDR + OFFSET[I];
  end
endmodule

module IRQ_ADDR2(input [7:0] IRQ, input [15:0] BASE_ADDR,
  output logic [15:0] ADDR);
  const logic [15:0] OFFSET [0 : 7] = {4, 8, 12, 16, 20, 24, 28, 32};
  logic [15:0] TEMP;
  always_comb begin
    TEMP = '0;
    for (int I = 0; I < 8; I++)
      if (IRQ[I]) TEMP = OFFSET[I];
      //This is NOT part of the "for" loop: Only one adder will be made
      ADDR = BASE_ADDR + TEMP;
  end
endmodule
```

■ **FIGURE 9.13** Minimizing logic in a loop prevents redundant circuitry from being created

The coding change in module of IRQ_ADDR2 in Figure 9.13 does a lot for circuit optimization, but the two circuit architectures are essentially the same. Architectural optimization can provide orders of magnitude differences in performance [5].

Finite impulse response (FIR) filters are a class of design where vastly different architectures can perform the same logic function. FIR filters have been extensively studied and many architectural implementations have been proposed. The following examples will present and contrast four architectures, two large and fast, the third small but slow, and a fourth in between in both area and speed.

All FIR filters satisfy the equation

$$y[n] = \sum_{k=0}^{N-1} bk * x[n-k]$$

where $y[n]$ represents the filter output, $x[n]$ represents the filter input, bk represents the filter coefficients, and N is the number of filter coefficients, that is, the order of the filter, or the number of taps. The convention of calling the order of a filter, its number of taps come from viewing the data input shift register of a filter as a delay line with outputs, or taps, at every multiplication stage, analogous to a water pipe with output taps spaced at regular intervals along its length.

This convolution equation may appear to be complicated, but in practice all it requires is a multiply accumulate function, which may be implemented in hardware or in software running on a standard microprocessor or specialized digital signal processor. The architectural tradeoffs are about deciding how many arithmetic units of what type to make and how to arrange data storage.

The required arithmetic can all be done with a single adder. On the other extreme, it can be done with N multipliers and $N - 1$ adders. In the latter case, one data sample can be processed per clock cycle. In the former, each sample will need many clock cycles, at a minimum one for every nonzero bit of the coefficients.

A diagram of a filter that can process a data sample every clock cycle is shown in Figure 9.14. This style is known as "direct" architecture. There is an M-bit wide shift register, N words deep for an order of N filter operating on M-bit data. In each clock

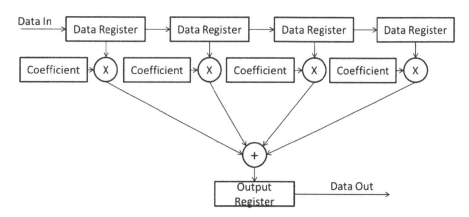

■ **FIGURE 9.14** Direct architecture FIR filter

cycle, a new *M*-bit wide data word is shifted in and oldest word is discarded. Each data sample is multiplied by a stored coefficient and the resulting products are summed.

The operating frequency of a direct architecture filter will be limited by the time it takes to multiply a data sample by its coefficient and then sum all the products. The direct architecure can be modified to build pipelining into it at the cost of increased register size: Instead of storing *M*-bit wide data samples, the transpose architecture shown in Figure 9.15 needs to store products and sums of products, with the required register size growing as more products are added to the cascade. However, the cycle time will be reduced to that of a multiply operation alone when a stage of pipelining is added after each multiplication or a multiply followed by a single addition without any extra pipelining.

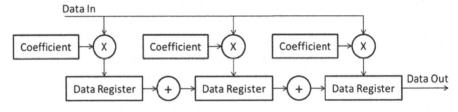

■ **FIGURE 9.15** Transpose architecture FIR filter

As the filters get large, direct architecture becomes an impractical implementation, as the final summation has too many terms to run in a reasonable amount of time. Transpose architecture can be expanded to any size, but a single source code file with hundreds of arithmetic operators may be impossible to synthesize with available computers. Creating a hierarchical transpose implementation is a solution to excessive computing power requirements. With hierarchical encoding, adders and multipliers of the appropriate size can be synthesized once and instantiated in the top-level design, minimizing the computing resources needed to synthesize large filters.

The same arithmetic functions can be done with nothing but a single adder. In Figure 9.16, a processor style filter is shown. It

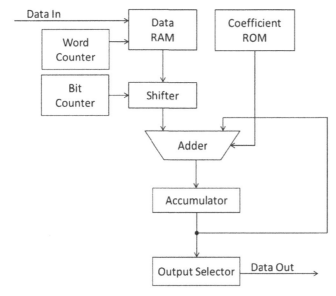

■ **FIGURE 9.16** Single adder processor architecture FIR filter

does have some hardware overhead not present in either of the preceding two design styles, meaning that for a very small filter, the processor style can be both bigger and slower than the preceding design styles, but its size will grow far more slowly than the others as the order of the filters increases. In Figure 9.16, an output selector is shown to allow the final output to be only the desired number of bits, which will usually be the same number of bits as the input data width. To keep full precision, the entire accumulator can be output instead.

A modification of the processor style filter is shown in Figure 9.17. Instead of using just a single adder and making multiple shifts and additions to perform each multiplication, it uses a hardware multiplier and an accumulator. It is slightly bigger than the first processor style filter but processes data faster, albeit at a slower clock rate.

While all architectures perform the exact same signal processing, performance can differ by orders of magnitude, far more than the variability that can be obtained by varying synthesis parameters

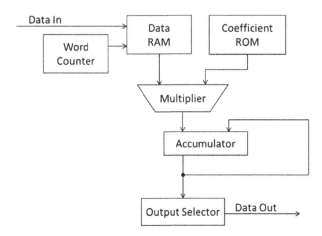

■ **FIGURE 9.17** Processor architecture FIR filter with hardware multiplier

or by minor coding technique changes. Table 9.1 gives area and speed data for representative examples of direct and processor style filters as width and depth are changed. These data are based on synthesis reports from a 90 nm standard cell library. Coefficients were pseudorandom, generated with a PERL (pattern extraction and recognition language) script, which is included in Chapter 12. Synthesis was performed without any constraints beyond specifying a clock that was too fast to be realized. Greater effort at optimization could be expected to improve both area and speed. All performance parameters given are for worst case operating conditions. More information on operating conditions will be given in Chapter 11.

In Table 9.1, the Filter Size column is width by depth, or the data sample size in bits by filter order. Area is given in square micrometers. Frequency is the maximum clock speed. Throughput is the rate at which data samples can be processed. For these implementations, transpose style filters always operate single cycle. Multiplier processor filters always take DEPTH cycles to process each sample. Processor style filters always take (WIDTH × DEPTH) clock cycles per sample. Throughput of processor style filters can be improved by skipping all cycles in which the coefficient bit is zero, at a cost of significantly more complicated controls.

Table 9.1 Performance and architecture

Filter Size	Architecture	Area	Frequency (MHz)	Throughput (MHz)
4 × 8	Processor	4,248	148	4.625
4 × 8	Mult. processor	5,002	136	17
4 × 8	Transpose	15,096	277	277
8 × 16	Processor	14,010	120	0.9375
8 × 16	Mult. processor	15,976	90	5.625
8 × 16	Transpose	88,549	200	200
12 × 24	Processor	23,034	92	0.32
12 × 24	Mult. processor	29,817	82	3.41
12 × 24	Transpose	225,270	146	146
16 × 32	Processor	47,805	31	0.06
16 × 32	Mult. processor	55,215	25	0.78
16 × 32	Transpose	474,777	133	133
20 × 36	Processor	53,478	26	0.036
20 × 36	Mult. processor	67,864	22	0.61
20 × 36	Transpose	778,614	121	121
24 × 40	Processor	68,992	25	0.026
24 × 40	Mult. processor	89,716	19	0.475
24 × 40	Transpose	1,186,718	108	108
28 × 48	Processor	92,240	21	0.015
28 × 48	Mult. processor	121,202	9.8	0.20
28 × 48	Transpose	1,897,965	108	108
32 × 64	Processor	147,770	23	0.011
32 × 64	Mult. processor	190,598	9.6	0.15
32 × 64	Transpose	3,121,180	99	99

The direct architecture, requiring as it does a combinatorial path with a cascade of DEPTH adders following a multiplier, is an impractical architecture for all but the smallest designs. Synthesis of such an impractical design would take more computing resources than may be available, while a hierarchical transpose format of the same design will quickly run on a modest computing platform and give far higher throughput.

Table 9.1 shows that the smallest but slowest design for a 32-bit, 64-tap filter can only process data at a rate of 11 kHz, and the transpose format filter can process data at 99 MHz, nearly 9000

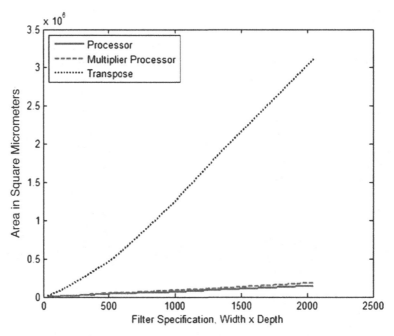

■ FIGURE 9.18 Silicon area as a function of arithmetic units

times as fast, although with an area 21 times as big. This sort of performance tradeoff can only be accomplished with architectural optimization, not by varying synthesis parameters or tweaking coding style. These same data are shown graphically in Figures 9.18 and 9.19. In the former, area for processor style filters, where the number of arithmetic units remains constant regardless of filter dimensions, is shown to increase far more slowly than that of parallel processing algorithms. In the latter graph, the price for small size is shown: parallel-processing architectures can be orders of magnitude faster than serial.

Filtering can also be done in software on a general-purpose processor. ARM assembly language code for a FIR filter is shown in Figure 9.20. Running on an NXP LPC3152 ARM9 processor at its maximum 180-MHz clocking rate, the loop would run in an even 100 ns. The LPC3152 is also constructed from 90 nm CMOS technology.

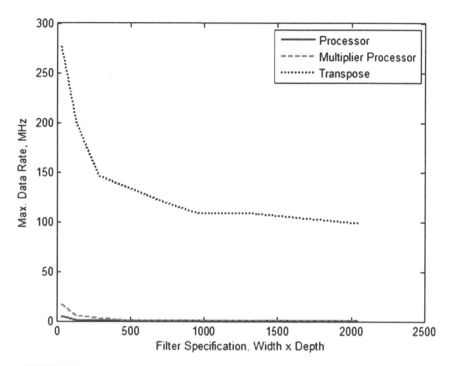

■ **FIGURE 9.19** Data processing speed as width and depth increase

LOOP	LDR	COEF, [COEF_ADDR], #4	;Fetch coefficient
	LDR	DATA, [DATA_ADDR], #4	;Fetch datum
	SMLAL	PROD0, PROD1, COEF, DAT	;Signed long multiply-accumulate
	STR	PROD0, [RESULT_ADDR], #4	;Store lower word
	STR	PROD1, [RESULT_ADDR], #4	;Store upper word
	SUBS	COUNT, COUNT, #1	;Decrement loop
	BNE	LOOP	;Repeat if not done

■ **FIGURE 9.20** ARM assembly language for a multiply accumulate loop

If the processor can be 100% dedicated to filtering with zero cycles for anything other than running the loop shown in Figure 9.20, a 32-tap filter function could then process data at 312 kHz, less than half the rate of the multiplier processor and over 400 times slower than the transpose architecture. Because 32-bit data with 64-bit products would risk overflowing the ARM processor's double precision capabilities, running the last example, a 64-tap filter with 32-bit data, would be impractical on that processor without compromising accuracy. The hardware algorithms have no such limitations. Their data paths can be made as wide and deep as desired.

Although a rather silly comparison, the area of the various filters can also be contrasted to that of a dedicated processor-integrated circuit. In a fine pitch ball grid array package, an LPC3152 processor takes 108 mm². The largest 16×32 filter would take substantially less than 1% of that much real estate while processing data over 400 times as fast.

Complete, scalable implementations of direct, transpose, hierarchical transpose, and both processor style filters are included in Chapter 12.

Filters are just one example of trading off operator reuse to optimize speed or area, one at the expense of the other. Another is shown in the code of Figure 9.21. In module "share_adder," only

```
module share_adder #(SIZE = 4) (input CLOCK, SEL,
  input [SIZE - 1 : 0] A, B, C, D,
  output logic [SIZE : 0] SUM);

  always_ff @(posedge CLOCK)
    if (SEL) SUM <= A + B;
    else SUM <= C + D;

endmodule

module dont_share_adder #(SIZE = 4) (input CLOCK, SEL,
  input [SIZE - 1 : 0] A, B, C, D,
  output logic [SIZE : 0] SUM);
  logic [SIZE : 0] SUM0, SUM1;

  always_ff @(posedge CLOCK)
   if (SEL) SUM <= SUM0;
   else SUM <= SUM1;

  always_comb SUM0 = A + B;
  always_comb SUM1 = C + D;
endmodule
```

■ **FIGURE 9.21** Resource sharing makes smaller but slower circuitry

one adder will be made. In module "dont_share_adder," two will be made. The module with two adders runs nearly twice as fast as the one with only one and takes nearly twice the area when synthesized.

POWER OPTIMIZATION

In the early days of integrated circuit design, the number of transistors that could fit onto a die was often the limiting factor on what could be produced. When a die could only hold 5000 transistors, it did not matter what else might be good about a design if it took 6000. If it could not be implemented in existing technology, the design was not useful.

Accordingly, synthesis tools in those days were set to aggressively pursue area optimization, with speed only optimized if it would not cause the area target to be missed.

With the exponential increase in transistor density that has been achieved in subsequent decades, area became plentiful and operating frequency became the most important design parameter. It would not matter what else was good about a 10-GHz networking card if it would only run at eight. This change in emphasis was reflected in synthesis tools, where the first priority became meeting frequency constraints and area was relegated to a secondary concern.

As transistors continued to shrink and ever more could be fit into a given area of a circuit, heat dissipation became a limiting factor in how fast a circuit could run. If the flipflops in a design can toggle at 20 GHz but having them all toggle at that speed would cause the device to overheat and fail, running at full speed all the time is not an option.

With the rise of portable electronics, power optimization to increase battery life has also become an important design consideration. Instead of featuring processor-operating speed as the primary measure of a device's comparative value, advertisements for today's consumer electronics are far more likely to prominently display battery life as a key selling characteristic. Long battery life depends on power optimization.

The following discussion of power dissipation in integrated circuits will assume that all circuits of interest are built from complementary metal oxide semiconductor (CMOS) technology. While this is the case for almost all current production semiconductors used for digital circuits, it has not always been true and may not remain true forever.

The central characteristic of a CMOS device is that it is built from pairs of N and P channel transistors, with one off when the other is on during normal operation. A pair of such transistors is shown in Figure 9.22. When used as a logic device, this pair will be an inverter, a logical NOT gate. A logic zero input will cause the top transistor to turn on and the bottom one to turn off. This will allow current to flow from the power rail to the output, but not to ground. In the opposite case, a logic one input will cause the bottom transistor to turn on and the top one to turn off, meaning that output will have a low impedance path to ground while the path from the power supply to the output will be cut off. A logic zero input becomes a logic one output and vice versa. More complex arrangements of such transistor pairs can be used to create any Boolean logic function.

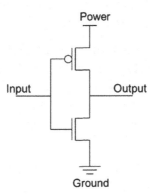

■ **FIGURE 9.22** CMOS transistor pair

In a CMOS circuit, loading is typically modeled as an ideal capacitor. While not perfect, this simplification gives adequately accurate results for most purposes. Going back to Ohm's law for power,

$$P = VI$$

and Ohm's law for current through a capacitor

$$i = C\frac{dV}{dt}$$

the instantaneous power consumed in a gate will be

$$P = CV\frac{dV}{dt}$$

When integrated over a cycle, the power consumed by a gate then becomes CV^2. Operated continuously at frequency f, the power consumed by a gate is

$$P = CV^2 f$$

In the general case, for N gates switching simultaneously at frequency f, each one having a dynamic parasitic capacitance of C_{pd} and using a voltage supply of V_{CC}, the capacitive component of transient power for a CMOS circuit is [6]

$$P_{TC} = C_{pd} \times V_{CC}^2 \times f \times N_{sw}$$

Although there are several different sources of parasitic capacitance, some internal to the gate and some external load, all can be lumped together to form C_{pd} when it comes to calculating overall power consumption.

In the ideal case, inputs would instantaneously switch back and forth between logic one and logic zero. The outputs would follow the inputs, also in zero time. In reality, inputs have rise and fall times. While the input is in an intermediate transitory state, both transistors can be partially turned on. With both transistors allowing some current flow, there is a path from power to ground. This transient current, also known as short-circuit current or crowbar current, is another component of dynamic power [7].

Short-circuit current will be proportional to the frequency at which the input changes. Unlike the capacitive component, short-circuit power is also proportional to input switching times. Total transient power is then the sum of the two transient components, or

$$P_{\mathrm{T}} = C_{\mathrm{pd}} \times V_{\mathrm{CC}}^2 \times f \times N_{\mathrm{SW}} + V_{\mathrm{CC}} \times I_{\mathrm{SC}} \times f \times N_{\mathrm{SW}}$$

where I_{SC} is the short-circuit current and the other terms are as in the capacitive transient power equation above. This equation assumes all gates have the same loading and short-circuit current, which implies that all transition times are identical. If that is too crude of an estimate, summation of individual gate values rather than multiplication by the number of gates switching would be necessary.

The other component of power consumption in a CMOS gate is static dissipation, a result of current leakage from power to ground. It is not frequency dependent. It is proportional to the overall size of the circuit and the power supply voltage.

In short, dynamic power is proportional to area and frequency. Static power is proportional to area.

Static power can be minimized by switching off the power to components that are not being used and using the lowest possible voltage for the entire design or block. The CMOS device of Figure 9.17 is modified in Figure 9.23 to add a power down transistor. When the power down signal is asserted, the logic circuit is put into sleep mode, consuming far less power than when in normal operating mode. Some design automation tools are capable of adding power down controls to individual gates or entire logic blocks, although this is not something that can be done in circuit description code. The power down transistor will be a high threshold device with lower leakage current than the normal logic transistors, which will be optimized for switching speed. Adding this type of power gating will add a significant area penalty.

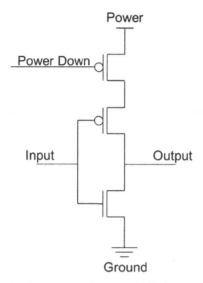

■ **FIGURE 9.23** Power down input and control transistor added to logic circuit

In years past, short circuit and static dissipation were so small in relation to the capacitive component of dynamic power dissipation that they could be ignored. As circuit geometries have shrunk, they have become far more significant.

In a clocked circuit, around 50% of the total dissipation will typically be in the clock distribution network. This is high in proportion to the area of clock networks compared to all the rest of the circuit, but clock networks by definition change state twice per period and random logic signals on average only change their values in a few percent of clock cycles.

One way to reduce power in the clock tree is to shut off branches of it on cycles when the active clock edge would not produce any state change. This technique, known as "clock gating," has shown average power reductions of 40% in some studies [8] although others report 20% as more realistic [9]. It is even less effective in field programmable gate arrays, where clock distribution networks are fixed.

Two versions of a flipflop circuit with an enable are shown in Figure 9.24. In the top version, feedback is used to maintain the stored value if the enable is false. In the clock-gating version at the bottom, the same function is accomplished via clock gating. The clock-gating circuit includes an active-low D latch to prevent glitches from the AND of the enable logic signal and the clock from creating spikes on the gated clock. A simple logical AND of the master clock and enable would result in short clock pulses and unreliable operation.

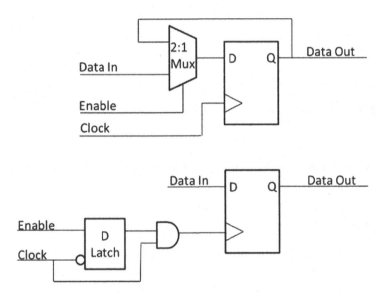

■ **FIGURE 9.24** Sequential feedback and gated clock flipflops

Since a D latch plus an AND gate are likely to be larger than a two to one multiplexor and the clock fan out doubles with clock gating added, for a single flipflop adding clock gating would increase both power consumption and area. A further disadvantage of the clock-gating design is that it introduces skew. The clock that arrives at the flipflop is delayed from the input clock and the flipflop output will be correspondingly skewed. Clock gating creates new clock domains. Reading a signal from the original clock domain into the new one could result in a timing violation. Going the other direction, from a gated clock flipflop to a flipflop clocked by the master clock, is less likely to present any problem, as the clock delay of the gated clock system will just add to the normal combinational and wiring delay of the path.

For a single flipflop, then, adding clock gating is entirely negative. It tends to be beneficial, at least from a power standpoint, when three or more flipflops operate with the same enable signal, although the added clock skew remains. Having three or more flipflops operating from the same enable is very common. An example of this would be a four-bit counter with enable, as shown in the code in Figure 9.25. By default, the clock would fan out to each flipflop and each flipflop would have a feedback path to hold state. When clock gating is used, the feedback paths and associated hardware are all eliminated and clock distribution stops at the clock-gating circuit in cycles when the count is not enabled.

The code of 9.25 could be explicitly modified to add clock gating, but modern synthesis tools can do that automatically when so instructed, although none will add it by default. Automating such tasks is highly recommended. Putting it into the code would likely result in errors and long debugging sessions.

```
module count_with_enable #(WIDTH = 4) (input CLK, RST, EN,
  output logic [WIDTH - 1 : 0] COUNT);

  always_ff @(posedge CLK, negedge RST)
    if (!RST) COUNT <= 0;
    else
      if (EN) COUNT <= COUNT + 1;
      else COUNT <= COUNT;
endmodule
```

■ **FIGURE 9.25** Scalable counter with enable, a good candidate for clock gating when WIDTH $>$ 2

Adding clock gating is not limited to pieces of a design where there is an explicit enable signal. XOR self-gating is a technique that can block clock propagation when there is no change on incoming data to a flipflop. In this arrangement, shown in

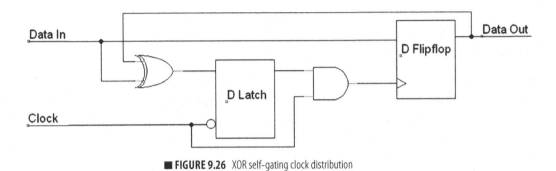

■ **FIGURE 9.26** XOR self-gating clock distribution

Figure 9.26, flipflop outputs are compared to incoming data via an XOR gate. If they are the same, the clock is blocked [10,11].

XOR clock gating can also be automatically inserted by synthesis tools.

Both these clock-blocking techniques turn a synchronous design into one that is something less than fully synchronous. Each of the gated clocks becomes a new clock domain, delayed from the master clock and skewed from each other. Without resynchronization, reliability of data transfers between gated domains will be compromised.

Clock gating can cause timing errors in a circuit that worked perfectly before gating was added. Despite violating the rules of synchronous design, it is often used in modern designs because of the power savings it offers.

Clock gating and minor code manipulation may gain some tens of percent improvement in power consumption, but architectural optimization can do far more. Again using FIR filters for architectural comparison, increasing functional units and decreasing clock speed will be shown to provide orders of magnitude decreases in dynamic power consumption.

Audio processing typically uses a minimum of 16-bit data with a 44.1-kHz sampling rate. Using a 40-tap low-pass filter with those parameters for an example, the difference between lowest and highest dynamic power was greater than a factor of 20. Because the power consumption for these designs at such low clocking rates is so insignificant and with the relative increase in leakage current with more advanced technologies, it would be a mistake to assume that a fully parallel solution is always best from a power standpoint. The transpose format design is more than a factor of 10 larger than the others, so when leakage power is added in, the total power consumption of the bigger design could become larger than that of the smallest even with the low clock speed. However, the story does not end there. Instead of operating continuously at the data rate, it may be possible to process data in parallel at a faster rate and then shut off power to the design until the next sample is needed. There is no single solution that can be blindly relied upon to always give the lowest power consumption.

Power consumption data for the 16-bit 44.1-kHz filter are shown in Table 9.2. In Table 9.2, clock speed is the rate at which it is necessary to clock the design to achieve a 44.1 kHz data throughput, not the maximum rate at which the synthesized circuits can run. All are capable of much higher clock speeds. Because dynamic power consumption is proportional to clock speed as well as area, operating at the slowest possible clock speed yields the lowest power consumption, despite the area penalty.

Table 9.2 Power consumption and architecture for a 16-bit, 40-tap filter

Architecture	Area	Clock Speed	Dynamic Power (μW)
Single adder processor	37,313	28.2 MHz	47
Multiplier processor	47,631	1.7 MHz	5.9
Transpose	505,575	44.1 kHz	2.16

A standard microprocessor could also be used to perform the filtering function. An LPC3152, running at its standard 180 MHz, would consume about 6 W: low for a general-purpose processor, but thousands of times higher than any of the custom solutions. This unfair comparison assumes that the custom-designed filters would fit into a small area of an integrated circuit that was going to be built anyhow and would not be a standalone device in its own package. It is beyond unlikely that building a small function like this as a complete integrated circuit performing no other functions would make sense from an economic or technical standpoint.

■ SUMMARY

The same source code often can be used to target both FPGAs and custom-mask ASICs, but there are some hardware considerations that influence coding style and limit design options.

Partitioning to keep combinational paths within a module yields better results than splitting paths across multiple modules. Registering module outputs and providing standardized data interchange protocols will help with design reliability and reuse.

Following the Rule of Seven helps make a design comprehensible at all levels of hierarchy. Keeping the intermodule communications to a minimum makes a design maintainable and simplifies debugging.

Design automation tools can move a design along the area/speed curve, but choosing the right architecture for design goals will have far more impact than anything that can be done by circuit optimization. Creating multiple functional units that operate in parallel can greatly increase operating speed at the expense of area.

Circuit techniques can also help with power reduction, but architectural choices again have far more impact than anything that can be done with gating. Clock gating also changes a synchronous design into one with many skewed clock domains, increasing the risk of timing errors.

If time to market is the only consideration, general-purpose processors can sometimes be substituted for circuit design, although the performance in both throughput and power consumption will be orders of magnitude better in a custom solution.

REFERENCES

[1] Design Compiler User Guide, Synopsys, 2006.

[2] Keating, M. "The Art of Good Design," Synopsys, 2009.

[3] Mike Keating, "The Art of Good Design," Synopsys, 2009.

[4] Evaluating Coding Styles, Synopsys CHIP Synthesis Workshop, 2002.

[5] Mehler RW, Zhou D. F.A.S.T. FIR Filter Synthesis. In: Proceedings of the 37th Southeastern Symposium on System Theory; Tuskegee, AL March 2005. pp. 320–5.

[6] CMOS Power Consumption and C_{pd} Calculation, Texas Instruments white paper, 1997.

[7] Piguet C. Low-Power CMOS Circuits: Technology, Logic Design and CAD Tools. CRC Press; 2005.

[8] White MA. Reducing Power with Advanced Synthesis. Synopsys Insight; 2011. Retrieved from http://www.synopsys.com/Company/Publications/SynopsysInsight/Pages/Art2-reduceadvsynthesis-IssQ4-11.aspx.

[9] Kommrusch S. Reducing Power in AMD Processor Core with RTL Clock Gating Analysis. EE Times; 2013. Retrieved from http://www.eetimes.com/document.asp?doc_id=1280395.

[10] Advanced Dynamic Power Reduction Techniques: XOR Self Gating, Synopsys white paper, 2011.

[11] Monteiro J, van Leuken R, editors. Integrated Circuit and System Design: Power and Timing Modeling, Optimization and Simulation: 19th International Workshop, PATMOS 2009. The Netherlands: Delft; 2009.

Chapter

Design for testability

Once a design has been completed, layout files are sent out for fabrication. Despite the best efforts of the foundry, not all the devices produced will be usable. Some will be defective. The bad ones should be screened out through testing.

A dense, pin-limited integrated circuit can be difficult to test. To facilitate testing, such circuits are almost always modified to add incorporate test circuitry.

Testing a device is not the same as verifying a design, although engineers frequently use "test" when they really mean verification. Verification ensures that the design is logically correct. Testing the physical circuits attempts to ensure that no defective devices are used in products.

FPGAs have testability built into them without any effort by the designer. FPGA designers do not need to modify their designs to make them testable.

YIELD, TESTING, AND DEFECT LEVEL

The yield of any manufacturing process – that is, the percentage that is usable – will never be 100%. With a new semiconductor process, the yield will start off closer to zero, rising as the process matures. To avoid putting defective components into products, a mechanism for differentiating the good parts from the bad is needed.

Soldering all the parts, good or bad, onto a board and determining later if some of the integrated circuits are defective because the systems they are in do not work is not a viable business plan. A rule of thumb is that the cost to find and fix a defective part rises by a factor of 10 at each level of design integration: if finding and rejecting a defective integrated circuit before use costs one, putting it onto a circuit board only to find out the board does not work because of the defective component costs 10. Putting the untested board into a system and finding out at that stage it does not work due to the defective chip costs 100, and shipping the whole untested system out to a customer and waiting for the customer to discover that it does not work costs 1000 [1].

In the earliest days of integrated circuits, the ratio of pins to gates was approximately one to one. An example of this would be a TTL 7400 package, invented by Texas Instruments and shown in Figure 10.1. This device has four NAND gates, and every single input and output is directly accessible from package pins.

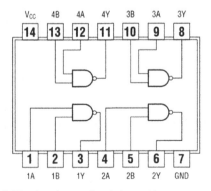

■ **FIGURE 10.1** TTL 7400 package (courtesy Texas Instruments)

Determining if the part works correctly is not challenging, as each gate can be fully controlled and observed from primary inputs and outputs.

The situation has changed drastically since such small-scale TTL components were state of the art. An integrated circuit package now can have over 2000 pins and tens of millions of gates, reducing the pin to gate ratio by four orders of magnitude. Determining if every internal node is correct with so little visibility into the circuit ranges from challenging to impossible.

Although determining that precisely 100% of a design has been manufactured correctly may not be possible, modifying the design to make it more testable usually will allow something acceptably close to 100% to be realized.

The foundry will attempt to produce only good devices. Once the devices have been produced, with some inevitable defect rate, engineers and technicians will test them in an attempt to differentiate the good parts from the bad. There too, inevitably some bad parts will slip through.

How many bad parts slip through the testing process is a function of how thoroughly the parts are tested. The rate at which bad parts are accepted as good is the defect level. It is the probability of accepting a bad part. The idealized version of testing is shown in Figure 10.2: New parts come in, they are tested, the good parts are accepted, and the bad parts are rejected.

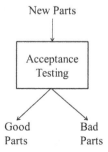

■ **FIGURE 10.2** Idealized testing, how it would work if fault coverage were 100%

Reality is somewhat less than ideal. Mixed in with the good parts will be some that are actually bad but still pass all acceptance tests.

If all possible manufacturing flaws are detected, the fault coverage is 1. If no effective testing at all is done, the fault coverage is 0. If the manufacturing process is perfect, the yield is 1. If no good parts are made, the yield is 0.

The relationship between these parameters is expressed in the Williams model [2], shown in the equation below. In it, D is the defect level, Y the yield, and T the fault coverage of the test protocols used.

$$D = [1 - Y^{(1-T)}] \times 100$$

From the Williams model, it is apparent that no matter the yield, if T is 1, the defect level is zero. This is never achieved. A defect level of a few hundred parts per million (PPM) is typically all that can be reached in a large, pin-limited design before the costs of modifying the design for testability and running the tests become prohibitive. How hard it is to approach a defect level of zero can be inferred by rearranging the Williams model to solve for the needed fault coverage with a given yield and defect level, as shown in the equation below.

$$T = 1 - \frac{\log(1-D)}{\log Y}$$

For a yield of 90%, which is typical of maturing technologies, achieving a defect level of 300 PPM would require fault coverage of over 99.7%.

For a newer process with a yield of only 50%, the fault coverage would have to go up to 99.957% to stay with a defect level of only 300 PPM.

Getting fault coverage of the last gates and flipflops can be difficult, leading to some temptation to cut short the generation of tests when the effort produces small and diminishing returns. The costs of doing so, however, can be substantial. If fault coverage only reaches 90% of the design in a process where the yield is 50%, the defect level goes up to 6.7% – 67,000 of every million parts accepted will actually be defective. A graph of defect level versus fault coverage as yield varies is shown in Figure 10.3. For the small ranges shown, the defect levels appear to be straight lines, but the exponential function defined by the Williams model is nonlinear.

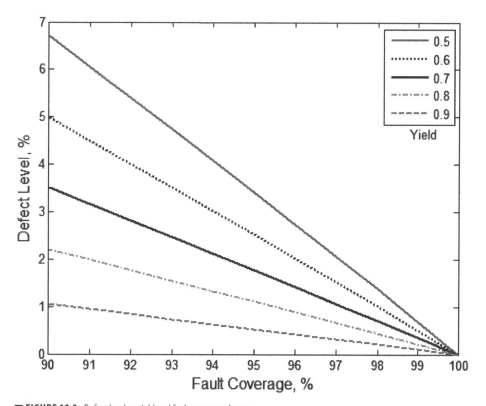

■ **FIGURE 10.3** Defect level as yield and fault coverage change

Figure 10.4 shows defect levels for a wider range of yields and fault coverage, going as low as 0.01% for yield and down to no-fault coverage at all. While such low levels of working parts would never be acceptable for regular production, new processes

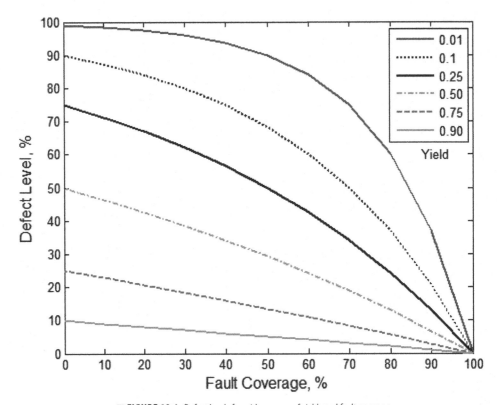

FIGURE 10.4 Defect levels for wider ranges of yields and fault coverage

in development can have very small levels of successful output. The graph also illustrates the importance of effective test. As the fault coverage goes down, the number of defective parts that are accepted and available to be put into products goes up.

When multiple components used on a board have some probability of being defective, the probability of having a working board declines exponentially with the number of potentially bad components, discounting any probability that the board itself is manufactured incorrectly.

If all components have an equal probability of being defective, the probability P of having a defective component on the board is

$$P = (1-D)^N$$

where D is the component defect level from the Williams model and N the number of components. The practical effect of this is that when multiple integrated circuits are put on a board, the probability of the board working is exponentially less than the probability of any individual component working. For example, to achieve a board defect level of 300 PPM on a board with 10 integrated circuits, each integrated circuit needs a defect level of about 30 PPM. Figure 10.5 shows how the expected board defect level varies with the number of components N with each component having a defect-level probability ranging from zero to 10%. The ranges of the graph are expanded beyond anything likely to occur in a production environment to emphasize the nonlinear nature of the relationships.

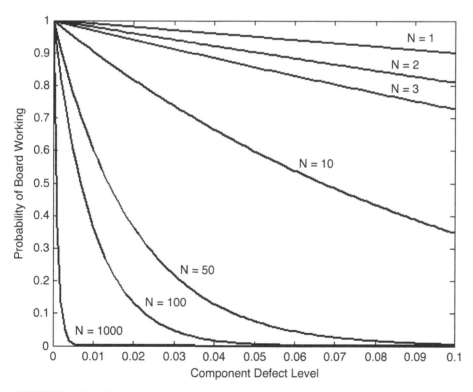

■ **FIGURE 10.5** Probability of a board working as number of components and component defect level rise

FAULT MODELING

There is no end to the manufacturing flaws that can lead to a defective integrated circuit. While the majority of flaws are short circuits in metallization layers and most of the rest are opens, or breaks in metallization, even if there were no other possible flaws, characterizing and devising a test for every possible flaw would be a never-ending quest.

To keep the number of tests down to a manageable number, abstract models that mimic effects of manufacturing flaws are employed.

The most common is the single stuck at (SSA) fault model. In this abstraction, each pin of each gate is assumed to be held to a constant logic zero or logic one, but only one line at a time is so faulted. The difference between a correct circuit and one with a single pin stuck at a constant value is the fault effect. If the fault effect can be seen at a primary output, then that fault, or any real manufacturing flaws that would cause similar behavior, is covered. This is the coverage that needs to approach 100% to achieve a defect level in the PPM range.

Inherent in the SSA fault model is the assumption that only one pin at a time will be faulted in the circuit under test. This unrealistic simplification is done entirely for reasons of computational complexity. The SSA fault model has twice as many faults as there are wires in the design. While not necessarily a small number, twice the number of pins is smaller than that produced by other fault models. If any number of lines can be simultaneously stuck, the number of stuck-at faults balloons to $3^N - 1$ for a circuit with N lines, since each node can be any of three states: Stuck at zero, stuck at one, or normal operation. Testing for so many possible fault conditions is unrealistic. Other fault models include transistor stuck on or stuck open, bridging faults causing wired AND or wired OR operation and delay faults. Computational complexity and lack of metrics for determining fault coverage limit the practicality of using these fault models, although they may be encountered.

A two-to-one multiplexor with internal node labels N1 through N3 is shown in Figure 10.6. To fully test it with the SSA fault model, each input, output, and internal node would be sequentially set to stuck at one and stuck at zero. A sequence of input vectors would then be applied. When there is at least one test that can distinguish between a faulted condition and the error-free case, all faults are covered.

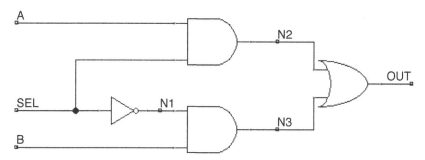

■ FIGURE 10.6 2:1 multiplexor with internal nodes labeled

An example of a test vector that covers a fault would be applying all ones to the inputs with A stuck at zero. A fault-free circuit would output a logic one, but the faulted model would output logic zero. Thus, the test vector A, B, SEL = {111} covers the fault A stuck at zero.

Since the design has three inputs, exhaustive testing would require 2^3, or eight, vectors. However, some vectors will detect multiple faults. In addition to detecting input A stuck at zero, input vector {111} will also detect the output stuck at zero and the output of the top AND gate stuck at zero. For this design, only four vectors are needed to detect all SSA faults. The four vectors and the faults they will detect are shown in Table 10.1.

Some vectors will detect multiple, independent faults. Another reason that only four vectors are needed to achieve 100% fault coverage for all 14 possible SSA faults of the multiplexor is that some faults are equivalent to others. With the SSA fault model, an AND gate input stuck at zero produces the same fault effect as

Table 10.1 Single stuck at fault detection for 2:1 multiplexor

Vector/Fault	100	111	010	011
A SA 0		X		
A SA 1				X
B SA 0			X	
B SA 1	X			
SEL SA 0				X
SEL SA 1	X		X	
N1 SA 0			X	
N1 SA 1				X
N2 SA 0		X		
N2 SA 1	X			X
N3 SA 0			X	
N3 SA 1	X			X
OUT SA 0		X	X	
OUT SA 1	X			X

the gate output stuck at zero. Table 10.2 shows fault equivalents for standard Boolean gates. Table 10.1 has no entries for XOR/XNOR gates, as they do not have any such equivalent faults.

Table 10.2 Fault equivalence

Gate	Fault	Equivalent To
AND	Any input SA 0	Output SA 0
OR	Any input SA 1	Output SA 1
NAND	Any input SA 0	Output SA 1
NOR	Any input SA 1	Output SA 0
NOT	Input SA 0	Output SA 1
NOT	Input SA 1	Output SA 0
BUF	Input SA 0	Output SA 0
BUF	Input SA 1	Output SA 1

Using fault equivalence to prune the test vector sets is commonly done to reduce test time. This technique, known as fault collapsing, seems to make sense but can actually reduce the number of bad parts that are screened out. The reason is that the SSA fault model does not accurately model all defective circuit behaviors and covering faults multiple times with different vectors increases the probability that defective operation will be detected.

With the vector set shown in Table 10.1, some faults will be detected multiple times. This partially compensates for the imperfect match between the SSA fault model and real faults in physical circuits.

Logically redundant gates present a particular problem for the SSA fault model, as they can be defective without changing circuit outputs. An example of this is the glitch-killing multiplexor first introduced in Chapter 2 and reproduced in Figure 10.7. Any of the faults that would result in the third input to the OR gate being stuck at zero could not be detected by the SSA fault model if the fault does not cause any other erroneous behavior.

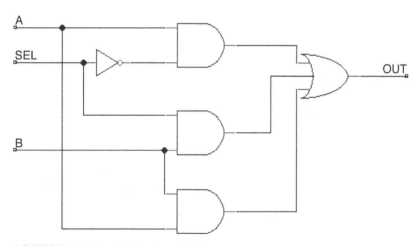

■ **FIGURE 10.7** Logic circuit with undetectable faults

Another limitation of the SSA fault model is shown in Figure 10.8. In that transistor-level model of a NOR gate, there is a break in the metallization that leaves the gate of one transistor floating [3]. This manufacturing flaw would not lead to an input or output being stuck at either logic level. There is no one vector that will detect the flaw, but a sequence of vectors can. If the vector {00} is first applied, the output will go to logic one, as it should. If the next vector is {01} or {11}, the output will go to logic one, again operating as it should. With this sequence of tests completed, it would be easy to conclude that the gate was operating properly. However, the sequence {00}, {10} would result in the output remaining at logic one for the second vector, at least until the charge from the first vector drained away. It would take this specific sequence for the fault to be detected with the SSA fault model.

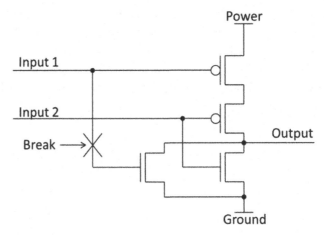

■ **FIGURE 10.8** NOR gate with a break in a line

Despite its limitations, the SSA fault model is used for testing most circuits, as it does detect a majority of error conditions, it can be run in an acceptable amount of time and it gives clear metrics for what has been tested and what has not. Because of its limitations, SSA testing is often supplemented with other types of tests, some of which will be covered later in this chapter. Use of fault models other than the SSA fault model would not change any of the design considerations in making a device testable.

ACTIVATION AND SENSITIZATION

For a fault to be detected, it must be active and a path to propagate the error to an external output must be sensitized. For the multiplexor of Figure 10.6, errors on input A could be activated by setting it stuck to either logic one or logic zero, but the faults could not be detected as long as SEL was logic zero. With SEL set to zero, no path is sensitive to A being stuck.

With a two-to-one multiplexor, activating all possible faults and sensitizing the output to each can be done with minimal effort. However, when many layers of logic and flipflops separate a gate from a primary output, activating a fault from a primary input, and then finding a path that can be sensitized for an error on the buried gate to propagate to a primary output can be difficult and require numerous cycles.

LOGIC SCAN

Because activating and sensitizing deeply buried faults is so difficult, logic circuits typically are modified to increase test access to internal nodes. The primary method for doing this is adding logic scan.

A digital circuit will typically have a structure similar to that shown in Figure 10.9. Data arrive on the primary inputs and pulse through the circuit, propagating through combinational logic between registers. The deeper the pipelines in the design, the more difficult it becomes to control and observe every node in the design, as is needed for testing. Designs with this structure are known as systolic arrays, as data pulsing through them with the clock are analogous to a heartbeat pumping blood through a living organism.

To reduce clutter, clock signals are not shown in the following diagrams. It is assumed in each case that there is one common clock distributed to each flipflop.

Adding logic scan to a design allows flipflops throughout the design to be set and read without going through any combinational logic. This facilitates checking deeply buried combinational gates. A systolic array modified for scan is shown in Figure 10.10. Modifying a design for logic scan involves adding a data selector, or two-to-one multiplexor, in front of each flipflop. When test mode

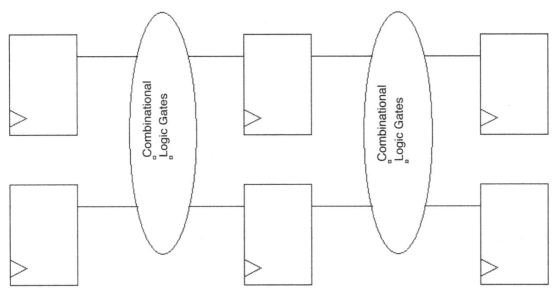

■ **FIGURE 10.9** Structure of a typical systolic array

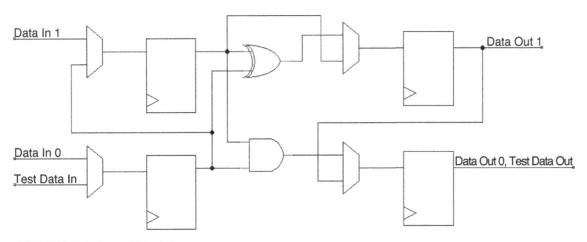

■ **FIGURE 10.10** Logic scan added to design

is enabled, Test Data In is selected as the input to the first flipflop in the chain.

With scan added, each flipflop can accept data from two different sources: the normal data path or test data via the scan chain.

When in scan mode, the entire design is one shift register. Serial test data are input from one pin and then pulse through the entire circuit as long as the circuit is left in test mode.

To test both the registers and combinational logic with scan, first a vector is threaded through the scan chain, setting up the initial conditions. Then, for just a single-clock cycle, the circuit is put back into normal operating mode. With logic scan disabled, the normal gate-level functions will be performed, resulting in the registers being updated.

Once the new data are clocked into the registers, test mode is again asserted. The modified data are shifted out for analysis. This procedure is repeated for each test vector.

An example of a piece of a systolic array that could be modified for test is shown in Figure 10.11. If the Data In signals were primary inputs, there would be no benefit derived from adding scan to the design, as the flipflops could be directly set from those inputs. However, if those flipflops were deeply buried, making them into part of a scan chain would allow them to be set to specific values without taking into account all the circuitry preceding them in the device.

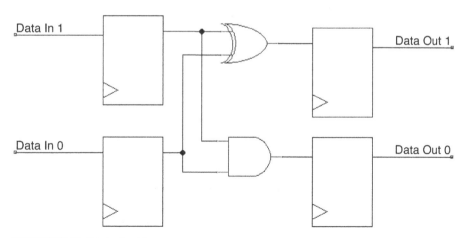

■ **FIGURE 10.11** Gate-level systolic array

In Figure 10.12, the gate-level design of Figure 10.11 is modified for logic scan. To reduce clutter, the test-enable signal is not shown. It would be connected to the select line of each multiplexor.

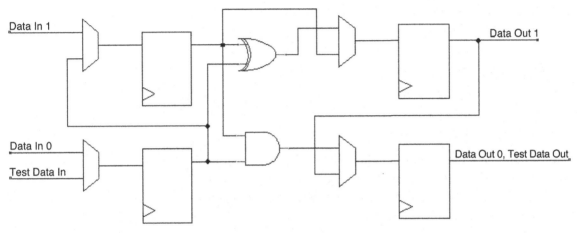

■ **FIGURE 10.12** Testing with logic scan

An example of a test vector for the scan chain of Figure 10.12 could be {XX11}. The first two bits are "don't care" for testing the two gates shown, as they will be propagated all the way to the output flipflops. The last two bits will set both input flipflops to logic one.

Once the vector is in place, test enable would be deasserted and the circuit clocked one time. If the circuit is working correctly, the top output flipflop would be set to logic zero and the bottom to logic one. The circuit would then be put back into test mode and the vector would be shifted out. If there is any discrepancy between the expected and received values, the circuit is deemed defective.

If that test passes, a second vector, perhaps {XX10}, could be shifted in. After this vector is shifted in and clocked in normal mode, the output flipflops should be opposite to their previous states. Again, if they are not, the circuit is defective.

With all the flipflops organized into a scan chain and enough vectors run through it, all SSA faults can be detected for a nonredundant systolic array circuit.

In practice, multiple scan chains will be created and run in parallel, as doing so reduces test time. For optimal test efficiency, the number of scan chains made should be the same as the number of parallel paths the test equipment has.

Modifying a circuit for scan is now highly automated. Instead of laboriously adding multiplexor functions, a modern synthesizer can replace all the flipflops with scan flops, which already have the extra test inputs built in. Some test implementations use dual-clock flipflops, with a separate clock distribution network for the test clock, or latch-based scan cells that require nonoverlapping two-phase test clocks.

While modifying a systolic array for scan insertion may be done automatically, not all designs are purely systolic arrays with a single clock. Some design techniques can make a circuit impossible to fully test and others can require modification to facilitate scan insertion.

Using an internal logic signal as a clock reduces test coverage. An example of this design technique is the ripple counter shown in Figure 10.13 [4]. It appears to offer some advantages: it is a

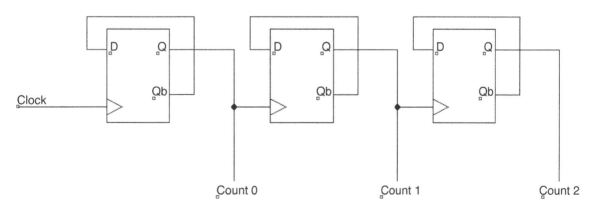

■ **FIGURE 10.13** Uncontrollable clocks leading to untestable components

binary counter but it does not require any gates, just flipflops. It would be small and fast.

Only the first flipflop would be controllable for scan. Even if a dedicated test clock is also routed to every flipflop, when switching back to normal mode the second and third flipflops would not advance in step with the other registers in the design. Any gates connected to the outputs of these flipflops could not be controlled during scan testing, reducing fault coverage.

Figure 10.14 shows a similar situation. With a logic signal driving an asynchronous input to a flipflop, again there is no way to directly control the state of the flipflop. The result would be reduced fault coverage.

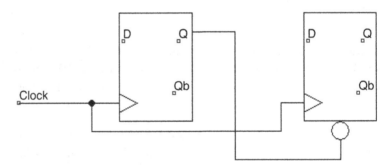

■ **FIGURE 10.14** Asynchronous reset controlled by a logic signal reduces test coverage

Both of these asynchronous design techniques need to be avoided for optimal fault coverage. Keeping designs fully synchronous maximizes test coverage and minimizes the number of defective components that will pass testing.

Internal Tri-state buses present another challenge for test. The issue with them is avoiding bus contention during testing. The problem is illustrated in Figure 10.15. While there is nothing wrong with the design, care must be taken in creating test vectors to ensure that there is always exactly one active driver. This can be a very computationally expensive exercise, leading to acceptance of a small vector set that does not achieve full fault

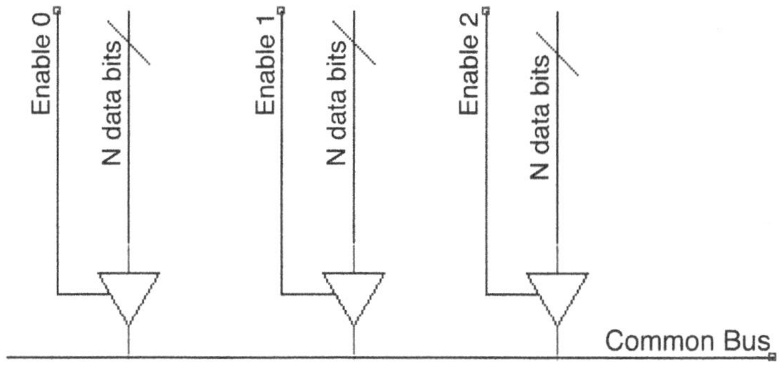

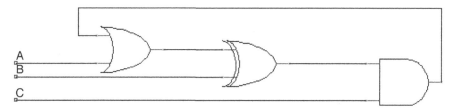

■ **FIGURE 10.15** Ensuring exactly one enable is always active is a challenge with test vectors

coverage. Minimizing the number of Tri-state buses in a design minimizes the problems associated with achieving high coverage without bus contention.

Combinational feedback loops may be simply prohibited by the semiconductor manufacturer. If they are used, they present a problem for test, as they can induce oscillation. An example of a combinational feedback loop is shown in Figure 10.16. The circuit shown will oscillate for A = 0, B = 1, and C = 1. The designer may know that this input combination will never occur in normal operation, but a test pattern generator can still use it.

■ **FIGURE 10.16** Combinational feedback loop

Some test pattern generators will detect such feedback loops and break them, at a cost of fault detection for the circuitry disconnected from the loop.

An awkward workaround for a design that must have combinational feedback would be to put a flipflop in that path that is multiplexed in only when in test mode, as shown in Figure 10.17. Adding this extra circuitry will cause additional delay in the loop during normal operation. Avoiding all combinational feedback is the best plan.

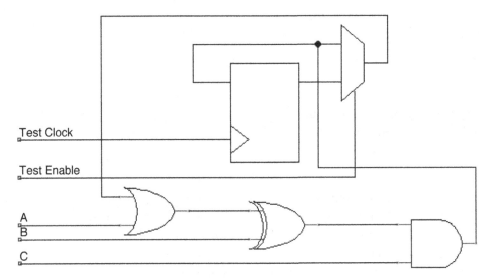

■ **FIGURE 10.17** Feedback loop broken with a flipflop for test

Gated clocks, introduced in Chapter 9, also may present a test problem, since they cannot be directly controlled from an external pin. Because of the skew they introduce to the clock distribution network, putting a gated clock flipflop in the same scan chain as an ungated one or one with a different gating structure can also lead to setup and hold violations when in scan mode.

To mitigate the controllability issue, the gated clock line can be multiplexed with a test clock, as shown in Figure 10.18. This cure has side effects, as it will increase the delay on both clocks, causing even greater skew between the gated clock flipflops and the rest of the design in both test and normal operating modes. Adding the multiplexor is unnecessary if the targeted library supports flipflops with a dedicated test clock input.

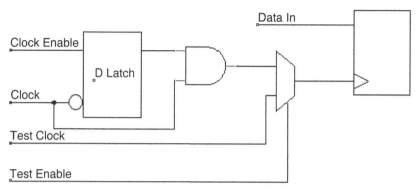

■ **FIGURE 10.18** Gated clock with test clock input

The clock gating circuit includes a latch. Use of latches in general reduces test coverage, as latches cannot be turned into scan cells like flipflops for most semiconductor technologies. The most common test procedure for latches is to hold them in their transparent state throughout testing. This leaves some lines untested. For maximum test coverage, latch-based design should be avoided.

BOUNDARY SCAN

Boundary scan, also known as JTAG (Joint Test Action Group, pronounced JAY-tag) and by its IEEE standard name, 1149.1, is similar to logic scan except that it only operates on top-level ports.

When boundary scan is inserted, each input and output is changed into a scan cell as shown in Figure 10.19. In normal operation, the scan cell adds the delay of a two-to-one multiplexor to each port but does not otherwise change functionality. In test mode, the entire perimeter of the design may be used as a shift register. One point on the device is connected to a port called TDI, or Test Data In. Another is connected to a TDO, or Test Data Out, port. All the other input and output cells are arranged such that the input from one cell is the output from another, until the entire perimeter of the design is included in one chain. Signals Update DR, Clock DR, Shift DR, and Mode are generated by the JTAG controller. Mode is not quite the same as the Test Mode Select (TMS) input

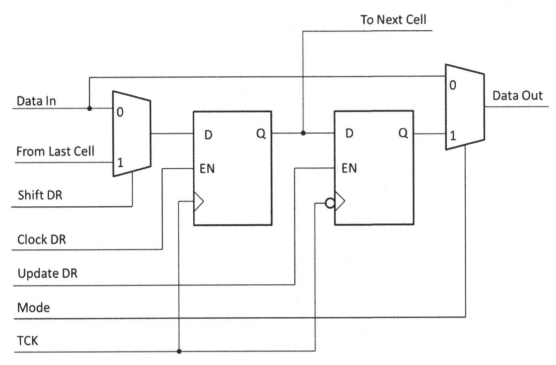

■ **FIGURE 10.19** JTAG boundary scan cell

shown in Table 10.3. When executing the BYPASS instruction, Mode must be logic zero to allow the device to operate normally. Otherwise, Mode will generally follow TMS.

Table 10.3 Test access port pins

Pin	Direction	Function
TDI	In	Test Data In
TDO	Out	Test Data Out
TMS	In	Test Mode Select
TCK	In	Test Clock
TRST	In	Test Reset (optional)

JTAG cells are frequently depicted with various control lines used as clocks. This is conceptually correct but does not represent best design practices. Using an enable signal or at least a gated clock as shown in Figure 9.22 will increase design reliability. Boundary scan registers (BSRs) have no special immunity to setup and hold violations. Following the rules of synchronous design is still advised. Other JTAG cells in this chapter are shown with logic signals used as clocks. This is done to simplify diagrams but does not reflect actual design practices, where they are typically used as flipflop enables or to gate the test clock.

In test mode, test data may be shifted into each scan device and throughout the entire system. Multiple JTAG-compliant components are typically arranged so that the test data output for one component is the test data input for the next, until the entire board or system is arranged into one chain, with one serial input and one serial output. The boundary scan chain of flipflops is called the BSR.

Boundary scan was invented to facilitate board, not integrated circuit, testing. By having a mode in which the pins of each device could be arranged into a shift register, it became possible to determine that connections to each component were correct and that each component was installed and at least minimally functional without need to probe the board.

Nothing prevents the scan data from being used for other purposes besides board test. They can be used for other internal tests and even configuring field programmable gate arrays or programming processors embedded in other devices. Almost all current digital designs incorporate JTAG.

Besides the BSR, a JTAG-compliant component requires a test access port (TAP), a state machine to control the various test signals, an instruction register (IR) and a bypass register. All must be added to the design. There are other, optional JTAG features such as a device identification (ID) register, an electronic chip ID register, an initialization data register (DR), and several status registers.

The TAP, also called the test access point, consists of inputs and outputs as shown in Table 10.3. All are required except test reset.

When pins are scarce, these signals may be multiplexed with other functions.

JTAG has three mandatory instructions, shown in Table 10.4. More instructions may be added, including user-defined instructions. In order to be able to decode at least the minimum instruction set, the IR must be at least two bits wide. The depth, that is, the number of instructions that may be loaded concurrently, must be at least two, so that a new instruction can be loaded while an older one is still valid. A new instruction is shifted into the active region of the IR in Update IR and Test Logic Reset states of the controller. Controller states are shown in Table 10.6 and Figure 10.21.

Table 10.4 Mandatory JTAG instructions

Instruction	Function
BYPASS	TDI and TDO are connected to each other via a single-bit bypass register.
EXTEST	The BSR is connected between TDI and TDO. The device acts as a shift register with data scanned in via the TDI port and out via TDO.
SAMPLE/PRELOAD	The BSR is connected between TDI and TDO. The device operates normally but test data may be scanned into the BSR.

The IR will typically be implemented as an N-bit wide shift register, where N is the ceiling $\log_2$ of the number of instructions implemented. The output stage of the shift register will be decoded to generate JTAG control signals. If the IR has more possible values than those needed for the implemented instructions, all extra combinations must be decoded as BYPASS.

A legal, minimal encoding for instructions is shown in Table 10.5. With only three instructions implemented, there must be two bit patterns that decode to BYPASS. No matter the IR width, a value of all ones must decode to BYPASS. The two least significant bits must be 01 for the SAMPLE/PRELOAD instruction.

Other common but optional JTAG instructions are HIGHZ, INTEST, and IDCODE.

Table 10.5 Minimal instruction register encoding

Bit Pattern	Instruction
00	EXTEST
01	SAMPLE/PRELOAD
10	BYPASS
11	BYPASS

HIGHZ places all the outputs other than TDO in high-impedance state and connects the bypass register between TDI and TDO. In this mode, data shift from TDI to TDO via the bypass register without changing any other outputs.

In INTEST mode, internal registers may be set and shifted out via TDI and TDO. This mode is similar to logic scan testing but is limited to a single logic scan chain.

In IDCODE mode, the optional ID register is shifted out via the TDO port least significant bit first.

One bit of an IR is shown in Figure 10.20. The IR must be at least two bits wide. In its minimal configuration, the "From Last Cell" input of the first bit would be connected to TDI and the "To Next Cell" output of the second bit would be connected to TDO.

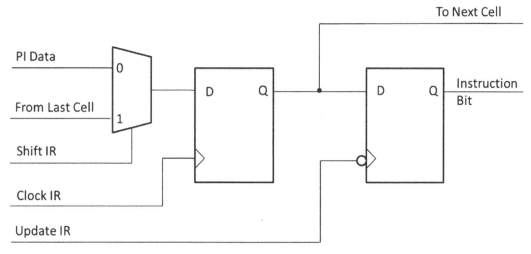

■ **FIGURE 10.20** One bit of a JTAG instruction register

The PI Data input is optional. It may be used to load other parallel test data from the device under test. It, and its associated two-to-one multiplexor, may be left out and the test data input connected directly to the D input of the first flipflop.

An asynchronous test reset may also be connected to IR flipflops. If a reset is used, some bits must be set rather than reset, as one of the JTAG requirements is that upon receipt of an asynchronous reset, the output of the IR must be set to IDCODE if that optional instruction has been implemented and to BYPASS otherwise.

When the IR updates normally, it does so on the falling edge of TCK. To facilitate this while violating the rules of synchronous design as little as possible, the logic signal Update IR may be used to gate TCK or as an enable with a negative edge-triggered flipflop. Clock IR too can be used as an enable for TCK rather than a clock signal to maintain synchronous operation.

So that operation may continue while new instructions are scanned in and to prevent a partially loaded instruction from sending out disruptive control signals, it is required that the IR have at least two flipflops per bit. The IR may be organized as a deeper shift register, with several intermediate stages between the final output stage, which is decoded to generate the JTAG control lines, and the input stage, which receives data from TDI. Figure 10.20 shows the minimal configuration, with only one shift stage before the final output stage.

The output of the IR is decoded in conjunction with the controller state to generate the JTAG control signals Shift DR, Update DR, Shift IR, Update IR, and any other signals needed to enable optional features.

If an ID register is implemented, it must have the structure shown in Figure 10.21. The least significant bit is always one. The next 11 bits are a manufacturer ID code, which is assigned by the IEEE. The following 16 bits are for the manufacturer's part number and the most significant four bits allow the manufacturer to add a version number for the part. An implementation of one bit of the ID register is shown in Figure 10.22.

31	28	27	12	11	1	0
Version		Part Number		Manufacturer ID		1

■ **FIGURE 10.21** JTAG ID register

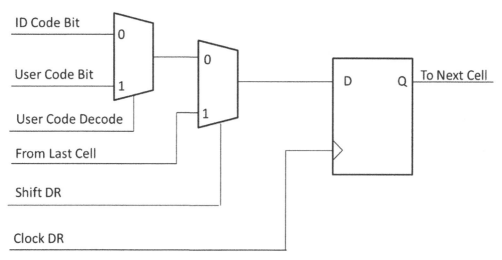

■ **FIGURE 10.22** JTAG ID register cell

When the IDCODE instruction is run, the 32 bits of the ID will be loaded into each ID register cell and scanned out via TDO. By requiring this instruction to be loaded at reset, the part can be identified at the start of testing. As with the other JTAG registers, the logic signal Clock DR would typically be used to enable the flipflops or gate the TCK rather than being actually connected to the clock input.

Another optional instruction is USERCODE. If this instruction is implemented, a second code is embedded in the device and can also be scanned out to TDO via the ID registers. This option is only used with programmable devices. When the USERCODE instruction is not implemented, there is no need for the first multiplexor and each ID code bit can be connected directly to the zero input of the only remaining multiplexor. The ID code bits would be hardwired to form the 32-bit ID value for the component.

JTAG-compliant devices must have a controller, which primarily consists of a 16-state state machine. States transition solely by reference to the TMS input, as shown in Table 10.6. State transitions are always on the rising edge of TCK. Figure 10.23 shows the transitions in a state diagram.

Table 10.6 JTAG controller state transitions

Current State	Next State, TMS = 0	Next State, TMS = 1
Test Logic Reset	Run-Test/Idle	Test Logic Reset
Run-Test/Idle	Run-Test/Idle	Select DR Scan
Select DR Scan	Capture DR	Select IR Scan
Capture DR	Shift DR	Exit 1 DR
Shift DR	Shift DR	Exit 1 DR
Exit 1 DR	Pause DR	Update DR
Pause DR	Pause DR	Exit 2 DR
Exit 2 DR	Update DR	Shift DR
Update DR	Run-Test/Idle	Select DR Scan
Select IR Scan	Capture IR	Test Logic Reset
Capture IR	Shift IR	Exit 1 IR
Shift IR	Shift IR	Update IR
Exit 1 IR	Pause IR	Update IR
Pause IR	Pause IR	Exit 2 IR
Exit 2 IR	Shift IR	Update IR
Update IR	Run-Test/Idle	Select DR Scan

When the 16-state controller is implemented as a four-bit binary state machine, every possible state is covered and all possible conditions for the state register represent a legal value. The state transitions are specified such that whatever the initial state of the flipflops, Test Logic Reset state will be reached by holding TMS high for five TCK cycles. This is why the Test Reset (TRST) input is optional. If the controller is implemented in some other manner, such as a one-hot machine, it may initially take on an illegal value and a reset would be necessary to commence normal operation. Needing the extra speed of a one-hot machine is highly unusual for a JTAG controller. A four-bit binary state machine is most commonly implemented.

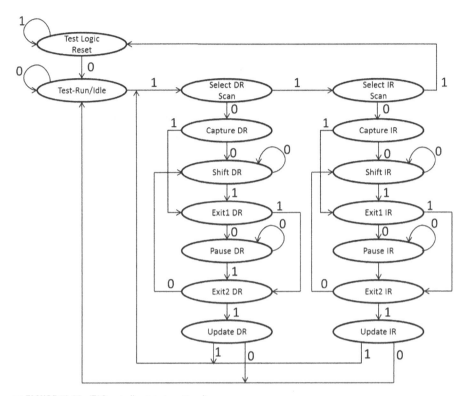

■ **FIGURE 10.23** JTAG controller state transition diagram

Once in the Test Logic Reset state, all test logic is disabled. The JTAG controller will remain in this state as long as TMS is held high.

If TMS is then set low, the controller transitions to Run-Test/Idle state. No JTAG action is required in this state, but design-specific testing may be enabled. Often it is used to enable some form of built in self-test (BIST). No other JTAG testing is enabled in this state.

Run-Test/Idle state will be maintained as long as TMS is held low. If it is again set high, one of two parallel paths may be chosen, although continuing to hold TMS high will result in returning to Test Logic Reset state after two more TCK cycles.

The two parallel paths operate on either the IR or DR. The DR in this context consists of the BSR, the Bypass Register, and the optional ID Register together.

Once one of the paths is chosen, the next state is Capture.

In Capture state, a word is parallel loaded into the lowest stage of the selected register on the rising edge of TCK. If operating in the IR path, the two least significant bits, which may be the only bits, must be 01. There is no such requirement for the DR.

In Shift state, the register is connected between TDI and TDO. A previously loaded pattern may be scanned out and a new pattern scanned in via TDI.

In Update state, previously loaded data are transferred into the output flipflops of the chosen register on the falling edge of TCK.

The Exit states provide methods of exiting the path or going back to Shift in the same path. Other JTAG registers are not changed in these states.

Figure 10.24 shows a design modified to include boundary scan. The core design is surrounded with boundary scan cells. In normal

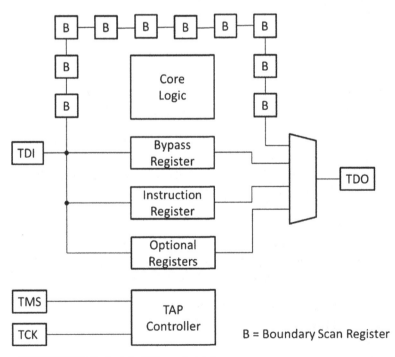

■ **FIGURE 10.24** Top level of a JTAG-compliant design

operation, the scan logic in these cells passively transmits data into and out of the core. It is only in test mode that they work as registers. The TAP controller, JTAG registers, and TAP inputs and output all need to be added to make a design JTAG-compliant.

Modifying a design for JTAG adds area, slows down paths, and requires pins. Although the performance penalties are substantial, JTAG is added to almost all digital integrated circuits to facilitate testing at the integrated circuit, board, and system levels.

Manually modifying a design to include JTAG testability is rarely necessary, as modern synthesis tools can automate the process. Field programmable gate arrays typically come with JTAG built in. Often they are configured via the TAP when the system is powered up.

BUILT IN SELF-TEST

Boundary and logic scan give external testers a view into the device under test. BIST adds circuitry so that the device can test itself and only report the results. BIST is most commonly used on embedded memories but also may be used to test random logic. BIST is typically enabled via a JTAG instruction.

Effective memory testing requires checking for more fault effects than logic testing. Some of the common memory faults are Address Decoder Faults, Stuck Open Faults, Coupling Faults, Disturb Faults, and Data Retention Faults.

Address Decoder Faults can manifest themselves in several different ways. For a given address, no cell might be accessed. A given cell might not be accessed for any address. An address might access several cells simultaneously. A cell might be addressed by several different addresses.

When a Stuck Open fault is present, a cell cannot be accessed due to a broken line. This may result in the previously accessed cell's data being redisplayed.

Coupling faults cause a victim cell to take on an erroneous value as a result of operation on an aggressor cell when the aggressor cell is accessed. The victim cell may take on the same value as the aggressor, the inverse of the value a value related to transitions in the aggressor cell.

Disturb faults cause an erroneous value to be set in a cell when the cell is read or when another cell is read.

Data retention faults allow charge to leak away, resulting in loss of stored data.

Given how many permutations are possible, exhaustive memory testing is not a serious option. A small subset of all possible tests is typically chosen. Writing a checkerboard pattern of zeros and ones is a common pattern, but will leave many errors undetected, including most address faults. Walking ones patterns and diagonal ones patterns have also been used. A test that will detect coupling faults is to set one cell to logic one and then surround it with cells containing zeros or the inverse and then accessing each cell. The cell being tested is then changed with each iteration through the test, which requires setting up and running the test for each cell. An address test involves writing sequential numbers to each address and then correctly reading back the entire memory array. This will guarantee to a high degree of probability that each address uniquely specifies one cell as long as the memory width is at least as great as the address bus size.

Each pattern will add some fault coverage at the cost of greater hardware overhead and test time. Figure 10.25 shows some commonly used memory test patterns applied to an eight-byte memory cell. Many others have also been used. Writing several patterns in sequence is often required to get an adequate level of fault coverage.

The general hardware strategy for BIST is to create a test pattern generator and a checker, as shown in Figure 10.26. When in self-test mode, data from a pattern generator are selected instead of normal operating data. When the device under test is a memory, the pattern generator will have to create addresses and control signals as well as test data.

Address	Pattern
0	00000001
1	00000010
2	00000100
3	00001000
4	00010000
5	00100000
6	01000000
7	10000000

Address	Pattern
0	01010101
1	10101010
2	01010101
3	10101010
4	01010101
5	10101010
6	01010101
7	10101010

Address	Pattern
0	00000000
1	00000001
2	00000010
3	00000011
4	00000100
5	00000101
6	00000110
7	00000111

■ **FIGURE 10.25** Memory test patterns. From top: walking ones, checkerboard, and address test

Memory self-test will typically use defined patterns such as those shown in Figure 10.25. Logic BIST will more often use pseudo-random patterns. Self-test pseudorandom pattern generators are usually formed with linear feedback shift registers (LFSRs). A three-stage LFSR is shown in Figure 10.27 and SystemVerilog code implementing it is shown in Figure 10.28. To avoid getting stuck at all zeros, the pattern generator needs to be initialized to something other than all zeros. An alternative that would also allow a maximal-length pattern to be generated would be to initialize it to all zeros but replace the XOR gate with an XNOR. Each

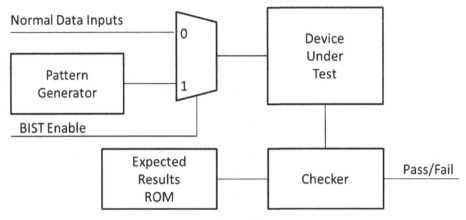

■ **FIGURE 10.26** General structure of BIST system

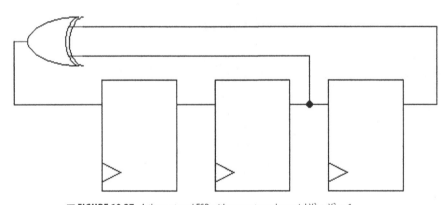

■ **FIGURE 10.27** A three-stage LFSR with generator polynomial $X^3 + X^2 + 1$

```
module LFSR #(SIZE = 3) (input CLK, RST,
output logic [SIZE - 1 : 0] PATTERN);
  always_ff @(posedge CLK, negedge RST)
    if (!RST) PATTERN <= '1;
    else begin
      PATTERN[0] <= PATTERN[SIZE - 1] ^ PATTERN[SIZE - 2];
      PATTERN[SIZE - 1 : 1] <= PATTERN[SIZE - 2 : 0];
    end
endmodule
```

■ **FIGURE 10.28** SystemVerilog implementation of an LFSR

stage of the LFSR can be tapped as an output to produce N-bit parallel data for an N-bit shift register, as is done in Figure 10.28.

Not all LFSR configurations will generate a maximal-length pattern of $2^n - 1$ for n flipflops. The test for determining if an LFSR will generate a maximal-length pattern is if the generator polynomial for the LFSR is prime. If it is prime, the sequence will be maximal. If it is not, the sequence will repeat after something less than $2^n - 1$ cycles.

LFSRs are defined by their generator polynomial, which is a mathematical expression of the terms that are fed back to the first flipflop of the shift register. For the LFSR of Figure 10.27, the generator polynomial is $X^3 + X^2 + 1$. If all three flipflops were fed back and XORed together to form the input to the first flipflop, the generator polynomial would be $X^3 + X^2 + X + 1$. This latter polynomial is nonprime, as using modulo two polynomial arithmetic, $(X^2 + 1) \times (X + 1) = X^3 + X^2 + X + 1$. The practical effect of being nonprime is that the output pattern will repeat with a different period depending on the initial seed, but all of the patterns will be shorter than maximal length. The final 1 in each polynomial is the input to the first flipflop and is always present, regardless of what feedback points are used.

The LFSR shown in Figure 10.27 is prime, or primitive, and will produce seven distinct patterns before repeating. While it is coded to be scalable, just changing the size while always feeding back the two most significant bits does not guarantee that the resulting pattern will be maximal length. An example of a nonprime polynomial of this format is $X^5 + X^4 + 1$. It can be factored to $(X^3 + X + 1) \times (X^2 + X + 1)$. Factoring modulo two polynomials requires using the rule that $(X^n + X^n = X^n - X^n = 0)$. There is no easy way to determine if a given polynomial is prime, but tables of prime polynomials of varying degrees have been created as well as software for generating such polynomials and checking if one is prime. Table 10.7 shows some primitive polynomials. There are more for each degree and the degree can be increased indefinitely. The reciprocal of each prime polynomial is also

Table 10.7 Some sample primitive polynomials

Degree	Polynomial
3	$X^3 + X + 1$
4	$X^4 + X^3 + 1$
5	$X^5 + X^3 + 1$
6	$X^6 + X^3 + 1$
7	$X^7 + X + 1$
8	$X^8 + X^4 + X^3 + X + 1$
12	$X^{12} + X^7 + X^4 + X^3 + 1$
16	$X^{16} + X^{14} + X^{13} + X^{11} + 1$

prime, so the examples of Table 10.7 can be used to find a second primitive polynomial of each length. To obtain the reciprocal of a polynomial, subtract the exponent from the degree of the polynomial for each term.

An example of converting a primitive polynomial to its reciprocal is shown below.

Original primitive polynomial: $X^3 + X + 1$
Subtract the exponent from the degree: $X^{3-3} + X^{3-1} + X^{3-0} = X^0 + X^2 + X^3$
Reciprocal primitive polynomial: $X^3 + X^2 + 1$.

The LFSR shown in Figure 10.27 is in external format, in that the XOR gate is external to the shift register. The same polynomials can form an internal LFSR as shown in Figure 10.29. System-Verilog code for that implementation is shown in Figure 10.30.

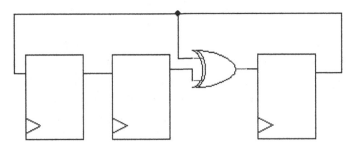

■ **FIGURE 10.29** Internal LFSR for polynomial $X^3 + X^2 + 1$

```
module LFSR_I #(SIZE = 3) (input CLK, RST,
output logic [SIZE - 1 : 0] PATTERN);
   always_ff @(posedge CLK, negedge RST)
     if (!RST) PATTERN <= '1;
     else begin
       PATTERN[0] <= PATTERN[SIZE - 1];
       PATTERN[SIZE - 1] <= PATTERN[SIZE - 2] ^ PATTERN[SIZE - 1] ;
       PATTERN[SIZE - 2 : 1] = PATTERN[SIZE - 3 : 0];
     end
endmodule
```

■ **FIGURE 10.30** SystemVerilog implementation of a scalable internal LFSR

The pattern produced by the two configurations will be different, but they will both cover the same range. If one implementation is prime, the other will be too. The pseudorandom patterns that the internal and external LFSRs for $X^3 + X^2 + 1$ will generate are shown in Table 10.8. The choice of seeding them with all ones was arbitrary. The same seven-value pattern will repeat no matter what the initial starting value is, as long as it is not all zeroes. If the device is reset to all zeroes, it will stay all zeroes forever.

Table 10.8 Patterns generated by internal and external three-bit LFSRs

Internal	External
111	111
011	110
110	100
001	001
010	010
100	101
101	011

The pattern generators used for BIST tend to be far longer than three bits and the test sequences used can be lengthy. While even a 32-bit LFSR does not need a lot of area, a ROM to check each of the four billion patterns one could generate would be a large cell. Accordingly, data compression and signature analysis are typically used with BIST.

The compression circuit and signature analyzer used with an LFSR tends to be almost an exact copy of the LFSR, although it need not be similar at all. One that could be used with the LFSR of Figure 10.27 is shown in Figure 10.31. SystemVerilog code implementing the circuit is shown in Figure 10.32. The only difference between the two circuits is that the receiver has an input for the data recovered from the circuit under test. The test procedure could be made more automated by adding a check for the correct signature and outputting only a single pass/fail line.

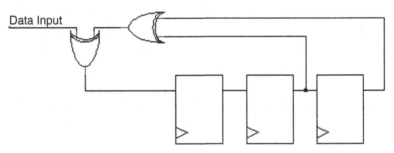

■ **FIGURE 10.31** Compression and signature circuit

```
module signature #(parameter SIZE = 3)
(input CLK, RST, DATA_IN, output logic [SIZE - 1 : 0] SIG);

always_ff @(posedge CLK, negedge RST)
  if (!RST) SIG <= 'b0;
  else begin
    SIG[0] <= SIG[SIZE - 1] ^ SIG[SIZE - 2] ^ DATA_IN;
    SIG[SIZE - 1 : 1] <= SIG[SIZE - 2 : 0];
  end
endmodule
```

■ **FIGURE 10.32** SystemVerilog code for compression and signature circuit

A BIST system with a signal generator and signature analyzer is shown in Figure 10.33. In that figure, parallel outputs are taken from the LFSR pattern generator but only a single line from the circuit under test is sent to the signature analyzer. For this to

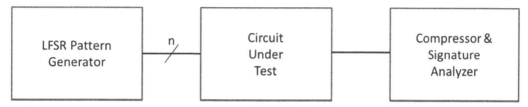

■ **FIGURE 10.33** BIST generator and signature analysis system

work, the circuit under test must have a structure something like that shown in Figure 10.34, with multiple gates fanning in to a single test point. While multiple signature analyzers may be run in parallel, adding one for every gate would be impractical.

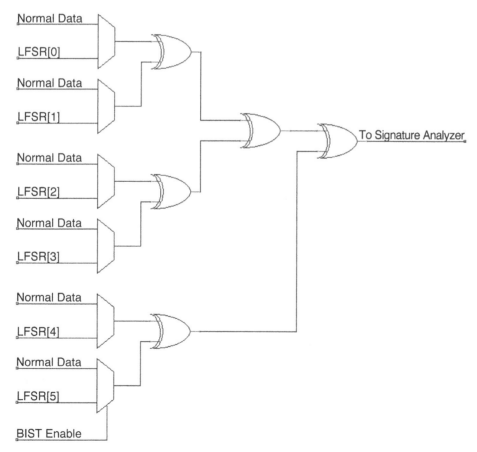

■ **FIGURE 10.34** A test point covering six data inputs; BIST enable would be connected to each multiplexor

To get around the problem of needing high fan-in nets, a signature analyzer can have multiple, parallel inputs. Figure 10.35 shows the circuit of Figure 10.31 modified to accept parallel data, turning an LFSR into a multiple-input LFSR, commonly shortened to multiple input shift register, or MISR. This arrangement can also be used for multiple high fan-in nets, increasing the fault coverage offered by each pattern generator.

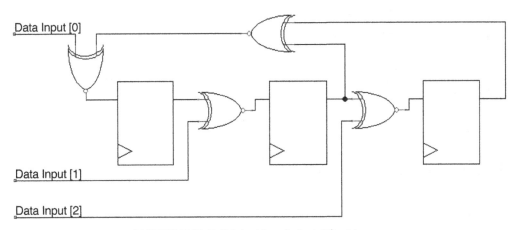

■ **FIGURE 10.35** Multiple input linear feedback shift register

Using an LFSR/signature analyzer system, there is some probability that a bad circuit will test good. This can happen because the pattern that will be left in the signature analyzer, being shorter than the pattern used to generate the signature, cannot be unique to a single generator pattern. For the three-bit circuit pair shown in Figures 10.27 and 10.31, the pattern generator will create a nonrepeating sequence of seven vectors. The circuit under test will then have 2^7 possible responses, but the compressor and signal analyzer can only end up in one of eight states. It is eight rather than seven because an all zeros signature is possible, although that is not a possible stimulus pattern. One of those eight will be the correct response, but there can be defective circuits that will also produce the same signature, even if the bit steam coming into the compressor is different from the fault-free bit stream.

The probability of failing to detect a defective circuit with this sort of a system is proportional to the length of the pattern used and the shift registers [5]. The general equation for the probability P of an aliasing error for a pseudorandom pattern of length L used with a signature analyzer shift register of length R is

$$P = \frac{2^{(L-R)} - 1}{2^L - 1}$$

For a system with a three-bit shift register using the maximal-length pattern of seven vectors, this comes out to a probability of failure of 11.8%, which would be unacceptable for any circuit. Increasing the shift registers to 16 bits gets the probability of aliasing error down to 0.016%, which may be acceptable. Since the cost of making an even longer set of registers is small, this sort of a system can be used to generate test patterns and signatures that are highly unlikely to fail due to aliasing.

These types of pseudorandom pattern generator and signature analyzer pairs are used for more than just BIST circuits. Communications systems use the same type of signature analysis to form cyclical redundancy checks that have a high probability of detecting data errors while keeping overhead low. They have also been used for encryption, but experienced code breakers would not take long to detect the type of pattern used and match the generator, enabling them to decrypt and read the hidden material.

BIST offers the advantage of operating at the full circuit speed. This allows delay faults that would be missed by external test systems to be detected.

Disadvantages of BIST are that adding it can substantially increase circuit area and delay paths. Unlike modifying a circuit for scan, which is a highly automated procedure, BIST circuits generally need to be crafted on an ad-hoc basis, although algorithms for memory BIST and pseudorandom pattern generators are well known and need not be developed anew for each implementation.

PARAMETRIC TESTING

Parametric testing involves powering up the device under test and seeing if the source current and voltages are within an acceptable range. Readings outside the set parameters typically mean the device has short circuits leading to excessive current draw or open circuits, leading to low or no current draw. Sometimes a tree of NAND gates is added to a device to measure input and output levels as well as delay times in crossing the die. The NAND tree can be multiplexed with other inputs and outputs to avoid dedicating any pins solely for this type of testing.

I_{DDQ} testing is a form of parametric testing in which the device is powered up but otherwise left in a quiescent state. The current draw is then checked to ensure that it is within the acceptable range, with outliers assumed to have opens or shorts. This class of testing generally does not require adding additional test circuitry.

■   SUMMARY

Yield is the percentage of parts that are correctly manufactured. Defect level is the percentage of defective parts that pass all tests. Getting the defect level down to an acceptable figure is the objective of testing.

Fault models are used to create a finite set of conditions that can be used to differentiate the performance of good parts from those with manufacturing flaws. The most commonly used such model is the SSA fault model, despite its limited performance. More comprehensive fault models can increase the time needed to test components to impractical levels.

Small-scale components give full access to all the internal gates, which makes testing them easy. As the ratio of pins to gates goes down, visibility into the circuit also goes down and testing becomes more difficult. To achieve a low defect level, large integrated circuits have incorporated additional circuitry to give better test access.

Almost all modern circuits have boundary scan added. Boundary scan was invented to facilitate board test but can also be used to test individual components. Similar in concept to boundary scan, circuits can be modified to add logic scan to further increase testability. The procedures for adding boundary and logic scan to a circuit are highly automated and require little manual modification of the design.

Self-test circuitry is also often added to circuits. It can be used for testing random logic but is more often used to test embedded memories. Tools for automating BIST are less common. Parametric testing can detect some types of manufacturing errors, sometimes without modifying the design.

REFERENCES

[1] IEEE Std 1149.1 (JTAG) Testability, Texas Instruments Semiconductor Group, 1997.
[2] Williams TW, Brown NC. Defect Level as a Function of Fault Coverage. IEEE Trans. Computers 1981;C-30(12):987–8.
[3] Garcia R. Rethink Fault Models for Submicron-IC Test. Test & Management World; October 2001.
[4] Jaramillo, K, Meiyappan, S. Philips Semiconductors "10 tips for successful scan design: part two," Electronic Design News, February 2000.
[5] Sebastian MJ. Application-Specific Integrated Circuits. Smith Addison-Wesley; 1997.

Chapter

Library modeling

OUTLINE

Once a design has been verified, the next step is logic synthesis. Logic synthesis produces a technology-dependent netlist implementation of the design. Such a netlist needs to include instances of cell models. All the cell models for a given target technology are gathered together into a library. This chapter deals with creating cell models for a component library.

COMPONENT LIBRARIES

The synthesizer will reference a library of components to map the design into the target technology. This library will typically contain hundreds of Boolean logic gates, complex cells, and flipflops. There may be numerous cells that are logically identical but have different speed and power characteristics.

Because gate-level simulation with annotated timing information is slow, designers may depend on timing reports generated by the logic synthesizer or other static timing analysis tools to ensure that the implementation meets all performance requirements, but simulating the netlist with the verification suite already developed often finds obscure timing problems not caught by static timing analysis. Before any such simulation can be done, a Verilog component library is needed. A component library is a link between an abstract behavioral design and the ultimate goal, a physical implementation.

Libraries have a cell model for each component. They will have to be updated for each new semiconductor process, even if no cells are added or subtracted from the library, as the performance characteristics of the cells will have changed.

An example of behavioral code for a four-bit counter and the synthesized netlist produced from it is shown in Figure 11.1. In this example, the Synopsys®, Inc. 90 nm SAED Generic Library [1] was targeted and the Synopsys Design Compiler tool was used for synthesis. This library is made available from Synopsys to universities for educational purposes and requires an educational addendum to the normal tool license [2].

```
module scalable_counter #(parameter SIZE = 4)
(input CLK, RST, output logic [SIZE - 1 : 0] COUNT);

always_ff @(posedge CLK, negedge RST)
  if (!RST) COUNT <= 0;
  else COUNT <= COUNT + 1;
endmodule

module scalable_counter ( CLK, RST, COUNT );
  output [3:0] COUNT;
  input CLK, RST;
  wire   N2, N3, N4, n1, n2, n5, n6, n7, n8;

  DFFARX1 \COUNT_reg[0] ( .D(n2), .CLK(CLK), .RSTB(RST), .Q(COUNT[0]), .QN(n2) );
  DFFARX1 \COUNT_reg[1] ( .D(N2), .CLK(CLK), .RSTB(RST), .Q(COUNT[1]), .QN(n8) );
  DFFARX1 \COUNT_reg[2] ( .D(N3), .CLK(CLK), .RSTB(RST), .Q(COUNT[2]), .QN(n1) );
  DFFARX1 \COUNT_reg[3] ( .D(N4), .CLK(CLK), .RSTB(RST), .Q(COUNT[3]), .QN(n7) );
  XOR2X1 U8 ( .IN1(n5), .IN2(n7), .Q(N4) );
  NAND2X0 U9 ( .IN1(n6), .IN2(COUNT[2]), .QN(n5) );
  XNOR2X1 U10 ( .IN1(n1), .IN2(n6), .Q(N3) );
  NOR2X0 U11 ( .IN1(n8), .IN2(n2), .QN(n6) );
  XOR2X1 U12 ( .IN1(n8), .IN2(n2), .Q(N2) );
endmodule
```

■ FIGURE 11.1 SystemVerilog behavioral code and synthesized netlist

In order to simulate the gate-level implementation, Verilog modules for the flipflops (cell DFFARX1) and each of the combinational gates would be needed. All the models for a given technology would be gathered together in one library. Some libraries have all cells in a single Verilog file and others have a separate file for each component, with the library then consisting of the directory where all the files are stored. The Synopsys library used in this example has all the cells in a single file.

CELL MODELS

Each cell model has both behavioral and timing data. For combinational cells, timing data are limited to propagation paths from inputs to outputs. Sequential cells also need timing checks to ensure that operating parameters such as setup and hold times are not violated.

In the simplest case, a cell model can have a single delay parameter added to the instantiation of a primitive component. Delay parameters can also be separated into rise times and fall times for more accurate modeling. Examples of such models are shown in Figure 11.2. This type of modeling is rarely used in modern

```
`timescale 1 ps/1ps
/*Simple model of a 2-input AND gate
with constant 1 picosecond delay*/
module AND21A(A, B, Q);
  input A, B;
  output Q;
  and #1(Q, A, B);
endmodule

/*2-input AND gate model with rise and
fall time parameters*/
module AND21B(A, B, Q);
  input A, B;
  output Q;
  //1 picosecond rise time, 2 picosecond fall time
  and #(1, 2)(Q, A, B);
endmodule
```

■ **FIGURE 11.2** Simple cell models with delays

libraries for timing simulation but can be found when only functional verification is being used. In Figure 11.2, the timescale directive at the top will be used for both modules. Having only a single timescale directive at the top of a file containing many library models is standard procedure.

A third field for turn-off time can be added for cells that have the capability of being put into a high-impedance state. Figure 11.3 shows a Tri-state® buffer cell that has a zero-to-one transition time of 1 ps, a one-to-zero transition time of 2 ps, and a turn-off time of 3 ps. With this type of delay modeling, the high impedance to zero time is the same as the one-to-zero time and the high impedance to one time is the same as the zero-to-one time.

```
`timescale 1ps/1ps
module TBUF1(IN, EN, Z);
   input IN, EN;
   output  Z;

   /*Rise, fall and turn off times*/
   bufif1 #(1, 2, 3) (Z,IN,EN);
endmodule
```

■ **FIGURE 11.3** Tri-state buffer model with rise, fall, and turn-off time parameters

A model of a more complex cell, such as an AND-OR-INVERT function, could have all the delay lumped together on the output driver or it could be distributed among the primitive operations. Both methods are shown in Figure 11.4. The two AND gates in each cell are in parallel, so the input-to-output delays of cells AOI1 and AOI2 are identical. The simple lumped delay model assumes all input to output paths have the same delay, which will not always be the case.

Paths from inputs to outputs can, as has been shown in Figure 11.3, have different delay characteristics for rise, fall, and turn-off times. In addition to these differences, semiconductors will not always perform the same under all operating conditions. Temperature, power supply, and manufacturing variability will

```
`timescale 1ps/1ps
/*Lumped delay AND-OR-INVERT cell*/
module AOI1(A, B, C, D, Q);
  input A, B, C, D;
  output Q;

  and (N1, A, B);
  and (N2, C, D);
  or  (N3, N1, N2);
  not #10 (Q, N3);
endmodule

/*Distributed delay AND-OR-INVERT cell*/
module AOI2(A, B, C, D, Q);
  input A, B, C, D;
  output Q;

  and #5 (N1, A, B);
  and #5 (N2, C, D);
  or  #4 (N3, N1, N2);
  not #1 (Q, N3);
endmodule
```

■ **FIGURE 11.4**   AND-OR-INVERT cell with distributed and lumped delays

affect the speed at which an electronic component can operate. Taken together, these three define a performance envelope for a component.

Each device will have a nominal operating voltage. However, it would be impractical to specify the voltage to infinite precision. Instead, a range is specified. For old TTL components, the supply voltage is 5 V, plus or minus a quarter of a Volt. More modern technologies use lower values but still have a range of acceptable voltage. Even if the supply voltage happens to be the nominal value to a thousand decimal places when the circuit is in its quiescent state, it will fluctuate with circuit switching, so some tolerance in the power supply is always needed.

Although circuits must continue to work as long as the power supply is within the specified range, they do not operate at exactly the same speed across the entire range of acceptable supply voltages. Operating speed decreases with reduced voltage.

Each device will also have a specified temperature range. Commercial components are normally specified to operate between 0 °C and 70 °C. Military grade components need to operate over a larger range, from −55 °C to +125 °C. As with supply voltage, delay characteristics will fluctuate with temperature. CMOS components tend to be linear with temperature for both rise and fall times, operating faster as the temperature drops. In TTL devices, fall times decrease with rising temperatures while rise times increase, but TTL is not used in modern integrated circuits.

In addition to these external variables, there will be some fluctuations in the delay characteristics of the devices produced when supply voltage and temperature are held constant. There are dozens of different factors that affect delays in semiconductors, but all can be lumped together to form a single "process" variable. Testing should weed out the devices that fail to meet specified delay characteristics, but there will remain some range of operating speeds for the devices that are deemed acceptable.

Process, temperature, and voltage can be combined to form scaling factors. When all three are optimal, the device will operate "best case." When all three are at their worst, a "worst case" scaling factor is applied to delays. Nominal delay values will form the "typical case" setting.

A semiconductor vendor may choose to provide separate libraries for each case or may have all in one library, with the delays multiplied by a scaling factor depending on which operating condition is chosen. When best, typical, and worst case delays are all in the same library, the rise, fall, and turn-off times of each cell are replaced by triplets for each field in the format Min:Typ:Max. Thus for a Tri-state cell, there will be three triplets, representing minimum, typical, and maximum delays for rise, fall, and turn-off times. Cells without Tri-state capability will only have two triplets, one for rise times and the other for fall times.

An example of a Tri-state buffer cell with three operating conditions is shown in Figure 11.5. When simulating worst-case conditions, the zero-to-one transition time will be 4.4 ps. For best case, the one-to-zero time will be 1.2 ps and for the typical case, the turn-off time will be 2 ps.

```
`timescale 1ps/1ps
module TBUF2a(IN, EN, Z);
  input IN, EN;
  output  Z;

  /*Rise, fall and turn-off times for
  best, typical and worst cases*/
  bufif1 #(1.6:3.2:4.4, 1.2:2.4:3.6, 1:2:3) (Z,IN,EN);
endmodule
```

■ **FIGURE 11.5** Tri-state cell model with best case, typical case, and worst case operating conditions

The distributed delay model offers the option of having some different paths through a cell, but its flexibility is still too limited for accurate timing simulation. Specify blocks are used to enumerate pin-to-pin delays across a cell.

The Tri-state buffer cell shown in Figure 11.5 has been modified in Figure 11.6 to use a specify block for timing data. Because that cell has only one data path from input to output, there is no

```
`timescale 1ps/1ps
module TBUF2b(IN, EN, Z);
  input IN, EN;
  output  Z;
  bufif1 (Z,IN,EN);
  specify
  /*Rise, fall and turn-off times for best, typical
  and worst cases*/
    (IN => Z) = (1.6:3.2:4.4, 1.2:2.4:3.6, 1:2:3);
  endspecify
endmodule
```

■ **FIGURE 11.6** Tri-state cell with delay data in a specify block

```
/*AND-OR-INVERT cell with specify block*/
module AOI3(A, B, C, D, Q);
  input A, B, C, D;
  output Q;

  and (N1, A, B);
  and (N2, C, D);
  or  (N3, N1, N2);
  not (Q, N3);
  specify
    (A, B *> Q) = 12, 10;
    (C, D *> Q) = 13, 11;
  endspecify
endmodule
```

■ **FIGURE 11.7** AND-OR-INVERT cell with a full connection specify block

obvious advantage to putting the delay information into a specify block, but not all cells are so simple.

In Figure 11.7, the AND-OR-INVERT cell first introduced in Figure 11.4 is shown with two different path delays from inputs to outputs. Use of the full connector operator *> indicates that both the path from A and the path from B have a 12 time unit rise time and a 10 time unit fall time. The full connector operator can also be applied to cells that have multiple outputs. The path specification shown below indicates that the paths from any bit of A to any bit of O will take 6.3 time units and that there are 64 total paths with the same delay characteristic.

(A[7:0] *> O[7:0]) = 6.3;

The above path specification indicates that any bit of A may cause a transition on any bit of O. The alternative for multiple input-to-output paths is to specify each in parallel, using => for each path. The following line of code indicates that there are eight parallel paths, each having a delay of 6.3.

(A[7:0] => O[7:0]) = 6.3;

Just as was done with parameters and localparams, giving meaningful, mnemonic names that can be updated in one place can help make cell models easier to understand and maintain. Within specify blocks, specify parameters, or specparams are used. Like localparams, specparams are fixed and cannot be redefined at compile time.

Figure 11.8 shows a two-input AND-gate model from the Synopsys library used in Figure 11.1. This model is written to allow two modes of operation: if "functional" is defined, a simple one-unit delay is used. When full timing information is required, rise and fall times from each input to the output are available. The specify block delays will override the unit delay associated with the primitive instance. There are separate libraries for best and worst case, so only one set of delays is included in each cell model.

```
module AND2X1 (IN1, IN2, Q);

output  Q;
input   IN1, IN2;

and #1 (Q, IN2, IN1);

`ifdef functional
`else
specify
  specparam in1_lh_q_lh=52,in1_hl_q_hl=50,in2_lh_q_lh=59,in2_hl_q_hl=56;
  (        IN1 +=> Q) = (in1_lh_q_lh,in1_hl_q_hl);
  (        IN2 +=> Q) = (in2_lh_q_lh,in2_hl_q_hl);
endspecify
`endif

endmodule
```

■ **FIGURE 11.8** AND cell with a specify block from Synopsys 90 nm library

The cell shown in Figure 11.8 uses +=> to specify that the path from the inputs to the output is noninverting and that a zero-to-one transition on an input will result in a zero-to-one transition on the output, if there is to be any change. Use of −=> in a specify block indicates an inverting path and a zero-to-one transition on an input that causes a change on the output should use the high-to-low delay figure rather than the low-to-high transition time. Plus and minus modifiers to delay paths can be used with full connections as well as with parallel paths.

An alternative syntax for specifying inverting and noninverting paths is to add a separate field for that in the path specification. The two lines below are equivalent.

```
(IN1 +=> Q) = 2, 1.5;
(IN1 => (Q +: IN1)) = 2, 1.5;
```

If a rising edge on IN1 would cause a high-to-low transition on Q, both of the following lines would indicate that the high-to-low transition time is to be used and a falling edge on IN1 would mean that the output delay would be that of a zero-to-one transition.

```
(IN1 −=> Q) = 2, 1.5;
(IN1 => (Q -: IN1)) = 2, 1.5;
```

Sometimes path polarity is conditional on the state of another input. An example of this is a two-input XNOR gate. If one input is steady state zero, then a change on the other one precipitates an inverting path from it to the output. Conversely, if an input is steady state one, a change on the other input would cause the output to follow the changing input with no inversion. To allow for such run-time path variations and always associate the right delay with each transition, conditional statements may be embedded in specify blocks. Unlike standard Verilog conditional statements, only "if" clauses are allowed. Neither "else if" nor "else" may appear in a specify block. A model for such a cell is shown in Figure 11.9. It also was extracted from the Synopsys 90-nm library.

USER-DEFINED PRIMITIVES

So far, all models used in these examples have had their functionality defined by the instantiation of a single built-in primitive or, in the case of the AND-OR-INVERT models, a small number of

```
module XNOR2X1 (IN1,IN2,Q);

output   Q;
input    IN1,IN2;

xnor #1 (Q,IN2,IN1);

`ifdef functional
`else
specify
  specparam in1_lh_q_hl=97,in1_hl_q_lh=114,in1_lh_q_lh=63,in1_hl_q_hl=64,
  in2_lh_q_hl=105,in2_lh_q_lh=78,in2_hl_q_lh=119,in2_hl_q_hl=80;
  if ((IN2==1'b0))
  (       IN1 -=> Q) = (in1_hl_q_lh,in1_lh_q_hl);
  if ((IN2==1'b1))
  (       IN1 +=> Q) = (in1_lh_q_lh,in1_hl_q_hl);
  if ((IN1==1'b0))
  (       IN2 -=> Q) = (in2_hl_q_lh,in2_lh_q_hl);
  if ((IN1==1'b1))
  (       IN2 +=> Q) = (in2_lh_q_lh,in2_hl_q_hl);
endspecify
`endif
endmodule
```

■ **FIGURE 11.9** Conditional paths in a specify block from Synopsys 90 nm library

primitives. While any function can be modeled with the built-in primitives, cells that use large numbers of them will simulate slowly.

To improve simulation performance, library developers may define their own primitive functions. User-defined primitives (UDPs) have their logical behavior specified in a table. Like built-in primitives, UDPs must have exactly one output. Combinational UDPs may have up to 10 inputs. Sequential UDPs are limited to nine inputs, as they also need a value for current state.

In addition to standard zero, one, and *x* values, UDP tables can use several symbols to decrease the number of table entries needed to fully specify the function. UDP symbols are shown in Table 11.1. Combinational UDPs only use the first two symbols. Sequential UDPs may use all.

Table 11.1 UDP table symbols

Symbol	Meaning
?	Iteration over 0, 1, and x
b	Iteration over 0 and 1
-	No change
r	Rising edge: 0 to 1 transition
f	Falling edge: 1 to 0 transition
p	Potential rising edge: 0 to 1, x to 1, or 0 to x transition
n	Potential falling edge: 1 to 0, x to 0, or 1 to x transition
*	Any transition

COMBINATIONAL CELLS

A full adder is a cell that could benefit from being implemented as a UDP. Since a full adder needs two outputs, two tables will be needed but this is in contrast to at least five primitive operators for a model without UDPs. Figure 11.10 shows a model for a full adder incorporating a module and two primitives.

The carryout term will be set to one whenever two out of the three inputs are one, so the third input in those cases is don't care. Use of a question mark for such cases covers any value for the third input, allowing the table to be shorter than if all values needed to be explicitly iterated. No such optimization is possible for the sum term, so all eight input combinations are enumerated there. In a UDP table, any unspecified input combination results in the output being set to unknown.

Question marks in tables are not the same as x values. In a table, an x would only match an x value on an input. It does not mean don't care, as it does with a casex statement. Any z values in a table are treated as x. The output of a UDP can never be high impedance.

While the outputs of the cell module are declared to be wires, declaring the outputs of combinational UDPs to be anything besides an output would be a syntax error. They are neither wires nor regs. This does not carry forward to sequential cells, where the outputs are registers.

```
module FULL_ADDER(A, B, CI, S, CO);
  input A, B, CI;
  output wire S, CO;
  ADDR_CARRY (CO, A, B, CI);
  ADDR_SUM (S, A, B, CI);
  specify
    (A, B, CI *> CO) = 3, 2.4;
    (A, B, CI *> S) = 2.5, 2;
  endspecify
endmodule

// FULL ADDER CARRY-OUT TERM
primitive ADDR_CARRY (CO, A, B, CI);
output CO;
input A, B, CI;
table
// A     B    CI  : CO
   0     0    ?   :  0;
   0     ?    0   :  0;
   ?     0    0   :  0;
   ?     1    1   :  1;
   1     ?    1   :  1;
   1     1    ?   :  1;
endtable
endprimitive

// FULL ADDER SUM TERM
primitive ADDR_SUM (S, A, B, CI);
output S;
input A, B, CI;
table
// A     B    CI  : S
   0     0    0   : 0;
   0     0    1   : 1;
   0     1    0   : 1;
   0     1    1   : 0;
   1     0    0   : 1;
   1     0    1   : 0;
   1     1    0   : 0;
   1     1    1   : 1;
endtable
endprimitive
```

■ **FIGURE 11.10** Full adder model with user-defined primitives

To model a complex cell that does have the ability to be put into high-impedance mode, the module would have to instantiate a Tri-state buffer as well as the UDP. An example of such a cell is shown in Figure 11.11.

```
module HiZ_Cell(A, B, EN, Z);
  input A, B, EN;
  output Z;

  //Q is output of UPD, not shown
  bufif1 (Z, Q, EN);
  Z_logic(Q, A, B);

  specify
    (A, B *> Z) = 2, 1.5, 1.7;
  endspecify
endmodule
```

■ **FIGURE 11.11** Cell model with high-impedance mode

A similar technique is used to form models of cells with complementary outputs. The UDP output can be connected to one output, but an inverter is then added to provide the complement. To keep the model symmetric, the noninverting output may use a buffer cell. A model of this type is shown in Figure 11.12.

```
module COMPLEMENTARY(A, B, Q, Qb);
  input A, B;
  output Q, Qb;

  //Z is output of UDP, not shown
  buf (Q, Z);
  not (Qb, Z);
  C_logic (Z, A, B);

  specify
    (A, B +=> Q) = 2, 1;
    (A, B -=>Qb) = 1.9, 0.9;
  endspecify
endmodule
```

■ **FIGURE 11.12** Cell model with complementary outputs

SEQUENTIAL CELLS

A sequential cell may be constructed using built-in primitive functions, but using a table for the logic will take fewer operators.

A D latch is the simplest sequential cell. A model for one is shown in Figure 11.13. All sequential cells can hold state and need a column dedicated to their current state. This latch is no exception, even though the behavior is never dependent on current state and all entries in that column are don't care. The model uses "–" to show that when the latch is not enabled, there is no change on the output for any value of the D input. Unlike combinational cells, the output of a sequential cell primitive is declared to be a register variable type, as it must have the capability to hold state.

Use of a primitive to define the logic of the cell allows the library developer to efficiently create many cells with identical logic but

```
module D_latch(D, ENA, Q);
  input ENA, D;
  output Q;
  latch (Q, D, ENA);
  specify
    (D +=> Q) = (1.5, 1.2);
    (ENA +=> Q) = (1.7, 1.4);
  endspecify
endmodule

primitive latch_logic (Q, D, ENA);
output Q;
reg Q;
input ENA, D;
table
  //  D    ENA   Present   Next
  //             state     state
      1    1    : ? :      1 ;
      0    1    : ? :      0 ;
      ?    0    : ? :      - ;
endtable
endprimitive
```

■ **FIGURE 11.13** D latch cell model

different timing. The primitive is instantiated in a module along with a specify block. It is in the specify block, and independent of the logic, that the timing characteristics of the cell are defined. The primitive follows the rules of all primitives, in that it has one output and the output is first in the port list. The module follows the general convention of Verilog modules in that the output is last in the port list.

A D latch, while sequential, is not edge sensitive. Flipflop models need edge behavior specified in their tables. Figure 11.14 shows a D flipflop cell using a table for behavior. In this model, edge transitions are specified using the notation (XY), representing a change from X to Y. The edge symbols shown in Table 11.1 can also be used.

```
module DFF(CLOCK, D, Q);
  input CLOCK, D;
  output Q;
  DFF_logic(Q, CLOCK, D);
endmodule

primitive DFF_logic(Q, CLOCK, DATA);
  output Q;
  input CLOCK, DATA;
  reg Q;
  table
//  CLOCK    DATA      Present    Next
//                     State      State
     (01)     1    :    ?    :     1  ;
     (01)     0    :    ?    :     0  ;
//Reduce pessimism for potential clocks
       p      1    :    1    :     1  ;
       p      0    :    0    :     0  ;
//No output change on clock falling edges
     (?0)     ?    :    ?    :     -  ;
//No output change with steady-state clock
       ?    (??)   :    ?    :     -  ;
  endtable
endprimitive
```

■ **FIGURE 11.14** D flipflop cell model

If any entry in a table uses edge constructs, the behavior for all edges on all inputs must also be defined. Failure to do this will cause the output to go unknown for any case not covered. The model of Figure 11.14 explicitly covers transitions on the D input, saying that as long as the clock is steady state, any transition on the other input should not cause any change on the output. It also needs to be specified that falling edges of the clock shall not cause any change on the output.

Should the clock transition from logic zero to unknown, the proper output depends on the current state of the flipflop. If the output is already in the same state as the input, there will be no change on the output if the clock goes to logic one or stays at zero. Those cases are covered in the table. However, there are no entries for the cases when the clock transitions from zero to unknown and the data input are not equal to the output. When that happens, the output will transition to X, which is the desired behavior. It does not need to be specified, as unspecified input combinations always result in the output being set to unknown.

Specify blocks for combinational logic only need to define the input-to-output delays. For sequential cells, those data need be supplemented with timing checks for run-time errors such as setup and hold violations. Timing checks for sequential models is shown in Table 11.2. Some simulators may not implement all timing checks, which will cause a model including an unimplemented check to fail.

Table 11.2 Sequential cell model timing checks

Timing Check	Relationship Tested
$setup	Data transition time before clock
$hold	Data transition time after clock
$setuphold	Data transition times before and after clock
$skew	Arrival times of two clock signals
$recovery	Time from reset release to clock arrival
$period	Clock period
$width	Clock duty cycle
$nochange	Stability between an edge and another signal

Not all flipflop models will use all the timing checks. $recovery is only applicable to cells that have asynchronous reset or preset inputs. $setuphold covers both setup and hold timing checks in a single line, so using it with either of the first two would be redundant. $nochange can be used for the same purposes as setup and hold checks. Specifying both a minimum high- and low-width time makes a period check redundant.

$skew reports a violation if the time between edges is greater than the specified limit. The other checks shown in Table 11.2 all report a violation if the difference between the specified events is less than the limit.

$skew is used to check that there is not too much clock skew between active edges of two clock branches. It reports a violation if the active edge of the second clock does not occur within a given amount of time from the active edge of the first clock. This timing check is not normally found in cell models.

Timing checks are used with a special register called a "notifier." Notifiers can be set to toggle when a timing check is violated. A notifier need not be initialized to any value to toggle. Verification code may be written to stop whenever a notifier changes state. Sequential UDPs will typically have an input for the notifier. When it toggles, the output is set to unknown. The D flipflop model of Figure 11.14 has been modified in Figure 11.15 to include a notifier. Notifier is not a keyword and may be used as the name of the notifier register. Any other legal Verilog identifier may be used as well.

The model shown in Figure 11.15 has two timing checks. A violation of either will cause the notifier to toggle and the cell output to be set to unknown.

Common timing checks and examples of their usage are shown in Figures 11.16–11.20. Checks may also have a second timing parameter specifying a threshold value to filter out spikes. The width timing check shown in Figure 11.20 is set so that a pulse shorter than the threshold value will not register as an error but

```
module DFF(CLOCK, D, Q);
  input CLOCK, D;
  output Q;
  reg NOTIFIER;
  DFF_logic(Q, CLOCK, D, NOTIFIER);
  specify
    $setup(D, posedge CLOCK, 1, NOTIFIER);
    $hold(posedge CLOCK, D, 0.5, NOTIFIER);
  endspecify
endmodule

primitive DFF_logic(Q, CLOCK, DATA, NOTIFIER);
  output Q;
  input CLOCK, DATA, NOTIFIER;
  reg Q;
  table
// CLOCK    DATA   Notifier      Present  Next
//                               State    State
    (01)     1       ?        :    ?   :   1  ;
    (01)     0       ?        :    ?   :   0  ;
//Reduce pessimism for potential clock edges
     p       1       ?        :    1   :   1  ;
     p       0       ?        :    0   :   0  ;
//No output change on clock falling edges
    (?0)     ?       ?        :    ?   :   -  ;
//No output change with steady-state clock
     ?      (??)     ?        :    ?   :   -  ;
//Set output unknown on any timing violation
     ?       ?       *        :    ?   :   x  ;
endtable
endprimitive
```

■ **FIGURE 11.15** D flipflop cell model with notifier to detect timing violations

one between the threshold and the width limit will. The threshold value in this case is the minimum width divided by 10.

Whether or not a timing check needs or can accept a threshold argument is implementation-dependent. There is considerable variation in the implementations of timing checks between design automation companies.

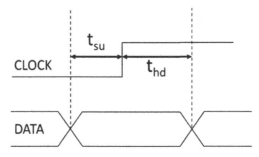

specparam tsu = 3, thd = 2;
$setup(DATA, posedge CLOCK, tsu, NOTIFIER);
$hold(posedge CLOCK, DATA, thd, NOTIFIER);
$setuphold(posedge CLOCK, DATA, tsu, thd, NOTIFIER);

■ **FIGURE 11.16** Setup, hold, and combined setuphold timing checks

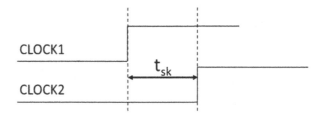

specparam tsk = 2;
$skew(posedge CLOCK1, posedge CLOCK2, tsk, NOTIFIER);

■ **FIGURE 11.17** Skew timing check

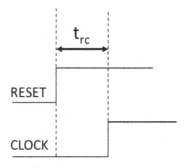

specparam trc = 2;
$recovery(posedge RESET, posedge CLOCK, trc, NOTIFIER);

■ **FIGURE 11.18** Recovery timing check

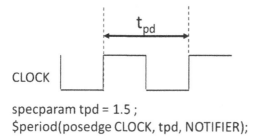

```
specparam tpd = 1.5 ;
$period(posedge CLOCK, tpd, NOTIFIER);
```

■ **FIGURE 11.19** Period timing check

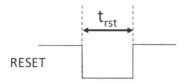

```
specparam trst = 3;
$width(negedge RESET, trst, trst/10, NOTIFIER);
```

■ **FIGURE 11.20** Width timing check

Flipflops that combine level and edge sensitivity add another level of complexity to models. Figure 11.21 shows a module for a model of a JK flipflop with active-low preset and reset inputs. The primitive for the model is shown in Figure 11.22.

The specify block for the JK flipflop model includes conditions on setup and hold checks. They prevent setup and hold violations on the J and K inputs from causing a timing violation when either the preset or reset is active. The triple & is used to indicate a logical AND only in specify blocks. Timing checks can also use equality operators, case equality operators, negation, and XOR operators.

The UDP table has no entry for both preset and reset being simultaneously active. When that happens, the output will go unknown. This is deliberate, as the behavior of such a flipflop is unpredictable when the two inputs are both active.

```
`timescale 1 ps / 1 ps
module JKFF(CLOCK, PRESET, RESET, J, K, Q);
  input CLOCK, PRESET, RESET, J, K;
  output Q;
  reg NOTIFIER;
  JKFF_logic(Q, CLOCK, J, K, PRESET, RESET, NOTIFIER);
  specify
    specparam tJSU = 12, tJHD = 6, tKSU = 12, tKHD = 6, tR = 2;
    specparam tPD = 20, tRST = 30, tW = 5;
    specparam tLH = 30, tHL = 27;
    $setuphold(posedge CLOCK &&& RESET, J, tJSU, tJHD, NOTIFIER);
    $setuphold(posedge CLOCK &&& RESET, K, tKSU, tKHD, NOTIFIER);
    $setuphold(posedge CLOCK &&& PRESET, J, tJSU, tJHD, NOTIFIER);
    $setuphold(posedge CLOCK &&& PRESET, K, tKSU, tKHD, NOTIFIER);
    $recovery(posedge RESET, posedge CLOCK, tR, NOTIFIER);
    $recovery(posedge PRESET, posedge CLOCK, tR, NOTIFIER);
    $period(posedge CLOCK, tPD, NOTIFIER);
    $width(posedge CLOCK, tW, tW/10, NOTIFIER);
    (J => Q) = tHL;
    (K => Q) = tLH;
    endspecify
endmodule
```

■ **FIGURE 11.21** JK flipflop model with asynchronous preset and reset inputs

```
primitive JKFF_logic (Q, CLOCK, J, K, PRESET, RESET, NOTIFIER);
  output Q;
  reg Q;
  input  CLOCK, J, K, PRESET, RESET, NOTIFIER;
  table
  // CLK  J   K   PRE RST NO :  Q :  Q+
      ?   ?   ?   0   1   ? :  ? :  1 ; //Async. preset
      ?   ?   ?   *   1   ? :  1 :  1 ;
      ?   ?   ?   1   0   ? :  ? :  0 ; //Async. reset
      ?   ?   ?   1   *   ? :  0 :  0 ;
      r   0   0   1   1   ? :  ? :  - ; //Hold on {JK} = 00
      r   0   1   1   1   ? :  ? :  0 ; //Reset on J = 0, K = 1
      r   1   0   1   1   ? :  ? :  1 ; //Set on J = 1, K = 0
      r   1   1   1   1   ? :  0 :  1 ; //Toggle on {JK} = 11
      r   1   1   1   1   ? :  1 :  0 ;
      f   ?   ?   ?   ?   ? :  ? :  - ; //Ignore negative edge of clock
      b   *   ?   ?   ?   ? :  ? :  - ; //Ignore data changes on steady clock
      b   ?   *   ?   ?   ? :  ? :  - ;
      ?   ?   ?   ?   ?   * :  ? :  x ; //Go unknown on any timing violation
  endtable
endprimitive
```

■ **FIGURE 11.22** UDP model for JK flipflop

MODEL PERFORMANCE

Chapter 2 emphasized the inefficiency of designing at the gate level. Up to this chapter, all subsequent coding had been done with behavioral constructs. Yet in modeling, primitive operators are back.

The reason for this is simulation performance. Models built from primitives, whether built-in or user-defined, simulate faster than behavioral code. The difference is significant. The D flipflop cell model of Figure 11.14 runs more than twice as fast as the behavioral model of a flipflop shown in Figure 11.23. Modifying the behavioral model of Figure 11.23 to set the output unknown for any timing violation would substantially complicate the model and further slow it down, increasing the advantage of using a table for such a cell.

```
module DFF_behav(CLOCK, D, Q);
  input CLOCK, D;
  output Q;
  reg Q;
  always @(posedge CLOCK) Q <= D;
endmodule
```

■ **FIGURE 11.23** Behavioral model of a simple D flipflop

■ SUMMARY

Once a design is verified, it is turned into a technology-dependent netlist through logic synthesis. This netlist may only be simulated if there are cell models for the targeted semiconductor process.

Cell models are built from primitive operators. These operators can be the built-in ones introduced in Chapter 2 or they can be user-defined cells.

Besides defining the behavior of a cell, a model will also define its timing. For combinational cells, this is just the propagation paths from inputs to outputs. Sequential cells also need timing checks to detect operating violations such as setup and hold times. When a timing violation is detected in simulation, the cell output is set to unknown.

Designing with primitive operators remains a tedious and error-prone exercise. Behavioral code is far more efficient for design. Once a design has been verified and synthesized, it may be resimulated to ensure proper timing. This requires a library of cell models that do use primitive operators along with timing data. The primitives used in cell models may be the built-in Boolean primitives or user-defined functions. Combinational cells are almost always constructed with the built-in primitives. Sequential cells almost always use a table to define behavior. Because a table can only have one output, tables are often supplemented with one or more instances of built-in primitives to complete a model of a complex cell.

REFERENCES

[1] Goldman R, Bartleson K, Wood T, Melikyan V, Babayan E. Synopsys' Low Power Design Educational Platform. In: Proceedings of the 9th European Workshop on Microelectronics Education (EWME 2012); Grenoble, France, 2012. pp. 23–6.

[2] "Synopsys 90 nm Generic Library for Teaching IC Design," http://www.synopsys. com/Community/UniversityProgram/Pages/Library.aspx.

Chapter

12

Design examples

This chapter shows implementations of designs referenced in earlier chapters. Included are complete, synthesizable designs for a state machine, several different filter designs, a FIFO and an asynchronous bit-serial interface. It also includes some test programs for the included designs.

STATE MACHINE

The following design example illustrates a canonical style of creating a design incorporating a state machine as described in Chapter 4. The state vector is latched into registers in a sequential code block. The state logic is created in a combinational block.

The design shown in Figure 12.1 is a sequence detector. It will set an "UNLOCK" output high for one cycle upon detecting the serial input sequence 1011, leftmost bit arriving first. To ensure that UNLOCK remains active for a full cycle regardless of any delay on the serial input data, UNLOCK is registered, which is a commonly accepted practice for module outputs.

This design example uses an enumerated type for the states, although that adds little if anything in this case over a design using binary encoding. The input sequence is required to be presented on successive clock cycles. A practical digital lock would be more complicated, in that it would have to detect pushing and

Digital Integrated Circuit Design Using Verilog and SystemVerilog   978-0-12-408059-1

```
module STATEMAC(input CLK, RST, SERIAL, output logic UNLOCK);
  input CLK, RST, SERIAL;
  //STATES is an enumerated type
  typedef enum{A, B, C, D} STATES;
  //STATE and NEXT_STATE are two variables of type STATES
  STATES STATE, NEXT_STATE;

  //Unlock when the sequence 1011 has been received.
  //Reset to zero to ensure it does not initialize to
  //unlocked.
  always_ff @(posedge CLK, negedge RST)
    if (!RST) UNLOCK <= 1'b0;
    else UNLOCK <= (STATE == D) && SERIAL;

  //State registers
  always_ff @(posedge CLK, negedge RST)
    if (!RST) STATE <= A;
    else STATE <= NEXT_STATE;

  //State logic
  always_comb
    case (STATE)
      A:  if (SERIAL) NEXT_STATE = B;
          else NEXT_STATE = A;
      B:  if (!SERIAL) NEXT_STATE = C;
          else NEXT_STATE = A;
      C:  if (SERIAL) NEXT_STATE = D;
          else NEXT_STATE = A;
      default:  NEXT_STATE = A;
    endcase
endmodule
```

■ FIGURE 12.1 State machine

releasing buttons, along with the complications of debouncing electromechanical switches and only advancing state when a valid button push was detected.

Two sequential blocks with the same sensitivity are used. This was done to emphasize the state machine structure. The UN-LOCK output could equally have been incorporated into the state vector assignment block. Both blocks are initialized with an asynchronous reset signal so that the device can be put into a known and benign state after power up.

In the simulation, the NEXT_STATE variable appears to change twice as fast as the STATE variable. This is because the serial input data arrive with the falling edge of the clock but the STATE variable can only change on the rising edge of the clock. Since the combinational circuitry of the state logic is sensitive to both the serial data input and the state, it can change values during the clock period. Rather than being a problem, this feature is illustrative of a benefit of using synchronous design. Combinational inputs may arrive at varying times during the period, but as long as they arrive with sufficient slack to meet the timing parameters of any flipflops in the design, the state vector and the output will remain stable for a full clock cycle.

There are numerous other ways to design this or any other sequence detector.

A small and noncomprehensive test fixture for the design of Figure 12.1 is shown in Figure 12.2.

```
module tb_STATEMAC;
  bit CLK, RST, SERIAL;
  STATEMAC  UUT(CLK, RST, SERIAL, UNLOCK);
  initial forever #1 CLK = ~CLK;
  initial begin
    RST = 1'b1;
    #2 RST = 1'b0;
    #2 RST = 1'b1;
    #2 SERIAL = 1'b1;
    #2 SERIAL = 1'b0;
    #2 SERIAL = 1'b1;
    #4 SERIAL = 1'b0;
    #2 SERIAL = 1'b1;
    #2 SERIAL = 1'b0;
    #2 SERIAL = 1'b1;
    #2 SERIAL = 1'b0;
  end
endmodule
```

■ **FIGURE 12.2** Test fixture for Figure 12.1

A waveform of the simulation of Figures 12.1 and 12.2 is shown in Figure 12.3.

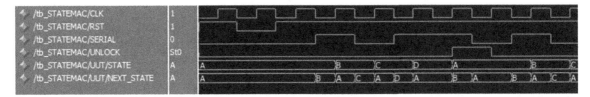

■ **FIGURE 12.3** Simulation of state machine code from Figure 12.1

FIR FILTERS

Figure 12.4 shows the simplest style of direct finite impulse response (FIR) filter. It is an implementation of the architecture shown in Figure 9.12. In this implementation, the output is of the

```
/*Direct architecture filter with data pipeline same depth
as order of filter. Output is same width as data input.*/

module direct #(WIDTH = 4, DEPTH = 4)
  (input CLOCK, input signed [WIDTH - 1 : 0] DATA_IN,
  output logic signed [WIDTH - 1 : 0] DATA_OUT);
  logic signed [WIDTH - 1 : 0] SR [DEPTH - 1 : 0];
  logic signed [2*WIDTH - 1 : 0] PRODUCTS [DEPTH - 1 : 0];
  parameter logic signed [WIDTH - 1 : 0] COE [DEPTH - 1 : 0] = {1,2,3,4};
  logic signed [2*WIDTH - 1 + $clog2(DEPTH) : 0] SUM;

  always_ff @(posedge CLOCK) begin
    SR[0] <= DATA_IN;
    SR[DEPTH - 1 : 1] <= SR[DEPTH - 2 : 0];
  end

  always_comb begin
    for (int I = 0; I < DEPTH; I++)
      PRODUCTS[I] = SR[I] * COE[I];
  end

  always_comb begin
    SUM = PRODUCTS[0];
    for (int I = 1; I < DEPTH; I++)
      SUM = SUM + PRODUCTS[I];
  end

  always_comb DATA_OUT = SUM[2*WIDTH - 1 + $clog2(DEPTH) -: WIDTH];
endmodule
```

■ **FIGURE 12.4** Direct mode FIR filter

same width as the input. Less-significant bits used in the calculation are truncated. The filter coefficients are assigned values when they are declared as parameters. For this demonstration, they are assigned the values of 1, 2, 3, and 4.

Figure 12.5 shows some slight variations on the design of Figure 12.4. In this implementation, the data pipeline is one word shorter and the first multiplication takes place on the incoming

```
/*Direct format filter with data pipeline one less than
filter order. Operates on input data. Output is full
precision of internal calculation, no truncation.*/

module direct2 #(WIDTH = 4, DEPTH = 4)
  (input CLOCK, input signed [WIDTH - 1 : 0] DATA_IN,
   output logic signed [2*WIDTH - 1 + $clog2(DEPTH) : 0] DATA_OUT);

  logic signed [WIDTH - 1 : 0] SR [DEPTH - 2 : 0];
  logic signed [2*WIDTH - 1 : 0] PRODUCTS [DEPTH - 1 : 0];
  parameter logic signed [WIDTH - 1 : 0] COE [DEPTH - 1 : 0] = {1,2,3,4};
  logic signed [2*WIDTH - 1 + $clog2(DEPTH) : 0] SUM;

  always_ff @(posedge CLOCK) begin
    SR[0] <= DATA_IN;
    SR[DEPTH - 2 : 1] <= SR[DEPTH - 3 : 0];
  end

  always_comb begin
    PRODUCTS[0] = DATA_IN * COE[0];
    for (int I = 1; I < DEPTH; I++)
      PRODUCTS[I] = SR[I - 1] * COE[I];
  end

  always_comb begin
    SUM = PRODUCTS[0];
    for (int I = 1; I < DEPTH; I++)
      SUM = SUM + PRODUCTS[I];
  end

  always_comb DATA_OUT = SUM;

endmodule
```

■ **FIGURE 12.5** Another variation on the direct mode filter data. Another variation is that there is no data truncation. All bits used to create the sum of products are output.

Figure 12.6 is the same as Figure 12.5 except the multipliers are in a separate module and are instantiated instead of inferred. This

```
/*Direct architecture filter with data pipeline one less
than order of filter. Operates on data input. Multipliers
are instantiated to allow hierarchical compilation.*/

module directH #(WIDTH = 4, DEPTH = 4)
  (input CLOCK, input signed [WIDTH - 1 : 0] DATA_IN,
  output logic signed [2 * WIDTH - 1 + $clog2(DEPTH) : 0] DATA_OUT);

  logic signed [WIDTH - 1 : 0] SR [DEPTH - 2 : 0];
  logic signed [2*WIDTH - 1 : 0] PRODUCTS [DEPTH - 1 : 0];
  parameter logic signed [WIDTH - 1 : 0] COE [DEPTH - 1 : 0] = {1,2,3,4};
  logic signed [2*WIDTH - 1 + $clog2(DEPTH) : 0] SUM;

  always_ff @(posedge CLOCK) begin
    SR[0] <= DATA_IN;
    SR[DEPTH - 2 : 1] <= SR[DEPTH - 3 : 0];
  end

  always_comb PRODUCTS[0] = DATA_IN * COE[0];

  generate
    genvar I;
    for (I = 1; I < DEPTH; I++)
      multiply #(WIDTH) M(SR[I-1], COE[I], PRODUCTS[I]);
  endgenerate

  always_comb begin
    SUM = PRODUCTS[0];
    for (int I = 1; I < DEPTH; I++)
      SUM = SUM + PRODUCTS[I];
  end

  always_comb DATA_OUT = SUM;
endmodule
```

■ **FIGURE 12.6** Direct mode hierarchical filter design

allows hierarchical compilation, which becomes necessary as the order of the filter becomes large. The instantiated multiplier is shown in Figure 12.7.

```
//Scalable multiplier

module multiply #(WIDTH = 4) (input signed [WIDTH - 1 : 0] OP1, OP2,
  output logic signed [2 * WIDTH - 1 : 0] PROD);

  always_comb PROD = OP1 * OP2;
endmodule
```

■ **FIGURE 12.7** Multiplier module referenced in Figure 12.6

A template of the transpose format architecture is shown in Figure 9.13. A scalable implementation of this architecture is shown in Figure 12.8.

```
module transpose #(WIDTH = 4, DEPTH = 4) (input CLK,
  input signed [WIDTH - 1 : 0] DATA_IN,
  output logic signed [2 * WIDTH - 1 + $clog2(DEPTH) : 0] DATA_OUT);
  parameter logic signed [WIDTH - 1 : 0] COE [0 : DEPTH - 1] = {1, 2, 3, 4};
  logic signed [2 * WIDTH - 1 : 0] PRODUCT [0 : DEPTH - 1];
  logic signed [2 * WIDTH - 1 + $clog2(DEPTH) : 0] PIPE[0 : DEPTH - 1];

  always_comb
  for (int I = 0; I < DEPTH; I++)
    PRODUCT[I] = DATA_IN * COE[I];

  always_ff @(posedge CLK) begin
    PIPE[0] <= PRODUCT[0];
    for (int I = 0; I < DEPTH; I++) PIPE[I + 1] <= PRODUCT[I + 1] + PIPE[I];
  end

  always_comb DATA_OUT = PIPE[DEPTH - 1];
endmodule
```

■ **FIGURE 12.8** Transpose architecture FIR filter

To allow large filters to be synthesized, the design of Figure 12.8 has been modified in Figure 12.9 to break out the arithmetic operations into separate modules that can be synthesized separately. In this design style, the internal operands grow with the stages of the filter, so the adders need to be larger than twice the input data size. Allowing for worst-case operation is accomplished by using the log ceiling function to increase the size of the adders. This design registers the outputs of both the multipliers and the adders. This is done to increase operating speed, although at the price of increasing area, power consumption, and latency.

```verilog
/*Hierarchical transpose format filter. Uses adder and multiplier module instances
that can be compiled separately.

All computed sum of products bits are output, meaning the output is more than
twice as many bits as the input. An alternative output assignment statement is
commented out. If the commented out version is to be used, the output statement
also needs to be changed to WIDTH - 1 bits.*/

module transposeH #(WIDTH = 4, DEPTH = 4) (input CLOCK,
  input signed [WIDTH - 1 : 0] DATA_IN,
  output logic signed [2 * WIDTH - 1 + $clog2(DEPTH) : 0] DATA_OUT);
  parameter logic signed [WIDTH - 1 : 0] COE [0 : DEPTH - 1] = {1, 2, 3, 4};
  logic signed [2 * WIDTH - 1 : 0] PRODUCT [0 : DEPTH - 1];
  logic signed [2 * WIDTH - 1 + $clog2(DEPTH) : 0] PIPE[0 : DEPTH - 1];

  always_ff @(posedge CLOCK)
    PIPE[0] <= PRODUCT[0];

  generate
    genvar I, J;
    for (I = 0; I < DEPTH; I++)  begin
      MULT  #(WIDTH) M(CLOCK, DATA_IN, COE[I], PRODUCT[I]);
    end
    for (J = 1; J < DEPTH; J++) begin
      ADD  #(WIDTH, DEPTH) A(CLOCK, PRODUCT[J], PIPE[J-1], PIPE[J]);
    end
  endgenerate

  always_comb DATA_OUT = PIPE[DEPTH - 1];
endmodule
```

■ **FIGURE 12.9** Hierarchical transpose format FIR filter

Figure 12.10 shows the registered-output adders and multipliers used in the hierarchical transpose format filter design of Figure 12.9.

```
//Scalable multiplier with registered output
module MULT #(WIDTH = 4) (input CLOCK, input signed [WIDTH - 1 : 0] DATA_IN, COE,
   output logic signed [2 * WIDTH - 1 : 0] PRODUCT);

   always_ff @(posedge CLOCK) PRODUCT <= DATA_IN * COE;

endmodule

//Scalable adder with registered output
module ADD #(WIDTH = 4, DEPTH = 4) (input CLOCK,
   input signed [2 * WIDTH - 1 : 0] OPERAND1,
   input signed [2 * WIDTH - 1 + $clog2(DEPTH) : 0] OPERAND2,
   output logic signed [2 * WIDTH - 1 + $clog2(DEPTH) : 0] SUM);

   always_ff @(posedge CLOCK) SUM <= OPERAND1 + OPERAND2;

endmodule
```

■ **FIGURE 12.10** Adders and multipliers referenced in hierarchical transpose design

Figure 12.11 shows a small test fixture for running all the filter models above in parallel. For this fixture, all the outputs were set to full precision rather than using truncation as is done in Figure 12.4. Because the number of pipeline stages is not consistent across all the implementations, there will be up to three clock cycles of skew between the outputs of the five filters. Implementations that operate directly on the input data may also have more output transitions than those that register the data inputs before use. For those implementations (direct and directH, Figures 12.5 and 12.6), the output data may not be valid for the entire clock cycle. This is a result of not registering outputs and the test fixture changing the data inputs coincident with the falling edge of the clock rather than the rising edge. Such design techniques are not acceptable in all working environments. Changing the final always_comb statement in those modules to always_ff @(posedge CLOCK) would register the outputs and eliminate spurious temporary output values at the cost of one more cycle of latency, larger area, and greater power consumption.

```
module tb_filter;
  parameter WIDTH = 4;
  parameter DEPTH = 4;
  reg CLK = 1'b0;
  reg signed [WIDTH - 1 : 0] DATA_IN;
  wire signed [2 * WIDTH - 1 + $clog2(DEPTH) : 0] DATA_OUT1,DATA_OUT2,
  DATA_OUT3, DATA_OUT4, DATA_OUT5;

  direct #(WIDTH, DEPTH) UUT1(CLK, DATA_IN, DATA_OUT1);
  direct2 #(WIDTH, DEPTH) UUT2(CLK, DATA_IN, DATA_OUT2);
  directH #(WIDTH, DEPTH) UUT3(CLK, DATA_IN, DATA_OUT3);
  transpose #(WIDTH, DEPTH) UUT4(CLK, DATA_IN, DATA_OUT4);
  transposeH #(WIDTH, DEPTH) UUT5(CLK, DATA_IN, DATA_OUT5);

  initial forever CLK = #1 ~CLK;

  initial begin
    DATA_IN = 1;
    #2 DATA_IN = 2;
    #2 DATA_IN = 3;
    #2 DATA_IN = 4;
    #2 DATA_IN = 5;
    #2 DATA_IN = 6;
    #2 DATA_IN = 7;
  end
endmodule
```

■ **FIGURE 12.11** Test fixture for direct and transpose filter architectures

An implementation of a scalable processor style filter as shown in Figure 9.14 is shown in Figure 12.12. It only uses one adder for all arithmetic functions, so it takes WIDTH × DEPTH clock cycles to process each data word. All the previous implementations operate single cycle.

A test fixture for the processor style filter is shown in Figure 12.13. Using the test data supplied, the output will always be zero, as only the most significant bits of the sum of products are connected to the output and small numbers are used for both coefficients and test data. The included debugging statements show that the correct values are being calculated.

```
module processor #(WIDTH = 4, DEPTH = 4)
  (input CLOCK, VALID, input signed [WIDTH - 1 : 0] DATA_IN,
  output logic signed [WIDTH - 1 : 0] DATA_OUT,
  output logic READY);

  parameter logic signed [WIDTH - 1 : 0] COE [DEPTH - 1 : 0] = {4,3,2,1};
  logic signed [2*WIDTH - 1 + $clog2(DEPTH) : 0] Accumulator;
  logic signed [WIDTH - 1 : 0] SR [DEPTH - 1 : 0];
  logic [$clog2(WIDTH) - 1 : 0] BITCOUNT;
  logic [$clog2(DEPTH) - 1 : 0] WORDCOUNT;
  logic START;

  always_ff @(posedge CLOCK)
    if (VALID) begin
      BITCOUNT <= 'b0;
      WORDCOUNT <= DEPTH - 1;
    end
    else begin
      if (BITCOUNT == WIDTH - 1) BITCOUNT <= 'b0;
      else BITCOUNT <= BITCOUNT + 1;
      if (BITCOUNT == WIDTH - 1 && WORDCOUNT > 0) WORDCOUNT <= WORDCOUNT - 1;
      else if (BITCOUNT == WIDTH - 1) WORDCOUNT <= WORDCOUNT - 1;
      else WORDCOUNT <= WORDCOUNT;
    end

  /*make shift register DEPTH deep to avoid having
  to operate on data input, which may not be stable
  for the entire processing period.*/
  always_ff @(posedge CLOCK) begin
    if (VALID) begin
      SR[0] <= DATA_IN;
      SR[DEPTH - 1 : 1] <= SR[DEPTH - 2 : 0];
    end
    else SR <= SR;
  end

  always_ff @(posedge CLOCK) begin
    START <= VALID;
    if (WORDCOUNT == 0 && BITCOUNT == WIDTH - 1) READY <= 1'b1;
    else READY <= 1'b0;
    if (START)
      Accumulator <= SR[DEPTH - 1] * COE[DEPTH - 1][0];
    else
      Accumulator <= Accumulator + ((SR[WORDCOUNT] << BITCOUNT) * COE[WORDCOUNT][BITCOUNT]);
  end

  /*Only output WIDTH most significant bits.*/
  always_comb DATA_OUT = Accumulator[2*WIDTH - 1 + $clog2(DEPTH) -: WIDTH];

endmodule
```

■ **FIGURE 12.12** Processor style filter

```
module tb_proc;
  timeunit 1us;
  timeprecision 1ns;
  parameter PERIOD = 1.0;
  parameter WIDTH = 4;
  parameter DEPTH = 4;
  reg CLOCK = 1'b0;
  reg VALID = 1'b1;
  reg signed [WIDTH - 1 : 0] DATA_IN = 1;
  wire signed [WIDTH - 1 : 0] DATA_OUT;
  wire READY;
  reg [1:0] DATA2 = 0;
  int EXPECTED;

  processor  UUT(CLOCK, VALID, DATA_IN, DATA_OUT, READY);

  initial forever CLOCK = #(PERIOD/2.0) ~CLOCK;

  always @(posedge CLOCK) begin
    //respond to UUT with a new word as soon as old word has been processed
    VALID <= READY;
    if (READY)  begin
      //send 1,2,3,4 repeating for data
      DATA2 <= DATA2 + 1; //will count 0 - 3
      DATA_IN <= DATA2 + 1; //will count 1 - 4
    end
  end

  always @(posedge READY) begin
    EXPECTED = UUT.SR[0] * UUT.COE[0] +
      UUT.SR[1] * UUT.COE[1] +
      UUT.SR[2] * UUT.COE[2] +
      UUT.SR[3] * UUT.COE[3];
      $display("SR Contents are %d, %d, %d, %d",
        UUT.SR[0], UUT.SR[1], UUT.SR[2], UUT.SR[3]);
      $display("Coefficients are %d, %d, %d, %d",
        UUT.COE[0], UUT.COE[1], UUT.COE[2], UUT.COE[3]);

    if (UUT.Accumulator != EXPECTED)
      $display("Error: Expected %d, Received %d", EXPECTED, UUT.Accumulator);
    else
      $display("Huzzah!:  Expected %d, Received %d", EXPECTED, UUT.Accumulator);

  end
endmodule
```

■ **FIGURE 12.13** Test fixture for processor style filter

Scalable code for the processor style with hardware multiplier filter of Figure 9.15 is shown in Figure 12.14. This one takes DEPTH clock cycles to process each data word. A test fixture for it is shown in Figure 12.15.

For data gathering, pseudorandom filter coefficients were used with varying filter widths and depths. The PERL script for generating the coefficients used is shown in Figure 12.16. While PERL is not a part of Verilog, many ASIC engineers find its use to be a convenient way of generating code and parsing computer-generated output such as synthesis reports.

When called, it needs to be supplied with two parameters, width and depth. Those numbers will be used to create a text file called COEWxD.txt, where W is the first parameter and D the second. If the file is saved as cogen.pl, the following command will yield a file called COE16x32.txt.

cogen.pl 16 32

Verilog has a built-in pseudorandom number generator $random that could be used in place of the PERL script of Figure 12.16. Figure 12.17 shows four different options for using the built-in generator. ALPHA will be a signed 32-bit quantity. BETA will also be a signed 32-bit quantity, but it is given an initial seed value so the sequence will be deterministic. It will give the same sequence each time it is run. GAMMA uses the modulus operator to limit the range of the numbers −9 to +9. DELTA uses the concatenation operator to force the numbers generated to be unsigned, effectively reducing the range to 0–9.

SystemVerilog has expanded pseudorandom number generation to include the tasks $urandom, which generates an unsigned number, and $urandom_range(X, Y), which generates an unsigned pseudorandom number between X and Y, inclusive. Both $urandom and $urandom_range can work with the SystemVerilog seeding task $srandom. Examples of $random, $urandom, and $urandom_range are shown in Figure 12.18. If $urandom_range is used with only one argument, the other argument will be assumed to be zero.

None of the random number tasks is synthesizable.

```
/*Processor style filter with pipelined multiply-accumulate.
READY signals to sending side that this filter can accept a
new word. STOP signals that the filter output is complete for
a given word.*/

module processorM #(WIDTH = 4, DEPTH = 4)
  (input CLOCK, VALID, input signed [WIDTH - 1 : 0] DATA_IN,
  output logic signed [WIDTH - 1 : 0] DATA_OUT,
  output logic READY, STOP);

  parameter logic signed [WIDTH - 1 : 0] COE [DEPTH - 1 : 0] = {4,3,2,1};
  logic signed [2*WIDTH - 1 + $clog2(DEPTH) : 0] Accumulator;
  logic signed [2*WIDTH - 1 : 0] PRODUCT;
  logic signed [WIDTH - 1 : 0] SR [DEPTH - 1 : 0];
  logic [$clog2(DEPTH) - 1 : 0] COUNT;
  logic START, NEWPROD;

  /*make shift register DEPTH deep to avoid having
  to operate on data input, which may not be stable
  for the entire processing period.*/
  always_ff @(posedge CLOCK) begin
    if (VALID) begin
      SR[0] <= DATA_IN;
      SR[DEPTH - 1 : 1] <= SR[DEPTH - 2 : 0];
    end
    else SR <= SR;
  end

  always_ff @(posedge CLOCK) PRODUCT <= SR[COUNT] * COE[COUNT];

  always_ff @(posedge CLOCK) begin
    START <= VALID;
    NEWPROD <= START;
    STOP <= READY;
    if (VALID) COUNT <= 0;
    else COUNT <= COUNT + 1;
    if (NEWPROD) Accumulator <= PRODUCT;
    else if (!STOP) Accumulator <= Accumulator + PRODUCT;
    else Accumulator <= Accumulator;
    //READY needs to go high on WORDCOUNT == DEPTH - 1 for max. performance
    if (COUNT == DEPTH - 1) READY <= 1'b1;
    else READY <= 1'b0;
  end

  /*Only output WIDTH most significant bits.*/
  always_comb DATA_OUT = Accumulator[2*WIDTH - 1 + $clog2(DEPTH) -: WIDTH];

endmodule
```

■ **FIGURE 12.14** Processor style filter with hardware multiplier

```
/*Test bench for processor style filter with hardware multiplier*/
module tb_processor_M;
  timeunit 1us;
  timeprecision 1ns;
  parameter PERIOD = 1.0;
  parameter WIDTH = 4;
  parameter DEPTH = 4;
  reg CLOCK = 1'b0;
  reg VALID = 1'b1;
  reg signed [WIDTH - 1 : 0] DATA_IN = 1;
  wire signed [WIDTH - 1 : 0] DATA_OUT;
  wire READY, STOP;
  reg [1:0] DATA2 = 0;
  int EXPECTED;

  processorM  UUT(.*);

  initial forever CLOCK = #(PERIOD/2.0) ~CLOCK;

  always @(posedge CLOCK) begin
    //respond to UUT with a new word as soon as old word has been processed
    VALID <= READY;
    if (READY)  begin
      //send 1,2,3,4 repeating for data
      DATA2 <= DATA2 + 1; //will count 0 - 3
      DATA_IN <= DATA2 + 1; //will count 1 - 4
    end
  end

  always @(posedge STOP) begin
    EXPECTED = UUT.SR[0] * UUT.COE[0] +
      UUT.SR[1] * UUT.COE[1] +
      UUT.SR[2] * UUT.COE[2] +
      UUT.SR[3] * UUT.COE[3];
      $display("SR Contents are %d, %d, %d, %d",
        UUT.SR[0], UUT.SR[1], UUT.SR[2], UUT.SR[3]);
      $display("Coefficients are %d, %d, %d, %d",
        UUT.COE[0], UUT.COE[1], UUT.COE[2], UUT.COE[3]);

    if (UUT.Accumulator != EXPECTED)
      $display("Error: Expected %d, Received %d", EXPECTED,UUT.Accumulator);
    else
      $display("Huzzah!:  Expected %d, Received %d", EXPECTED,UUT.Accumulator);

  end

endmodule
```

■ **FIGURE 12.15** Test fixture for hardware multiplier processor filter

```
#!/usr/bin/perl
use warnings;
use strict;

if($#ARGV != 1) {
    print STDERR "You must specify width and depth.\n";
        exit 4;
}

my $filename = "COE" . $ARGV[0] . "x" . $ARGV[1] . ".txt";

open (OUTFILE, ">$filename") or die "cannot open output file \n";

print OUTFILE "{";
for (my $I = 0; $I < $ARGV[1]; ++$I) {
        print OUTFILE $ARGV[0];
        print OUTFILE "'d";
        print OUTFILE int rand(2**$ARGV[0]);
        if ($I < $ARGV[1] - 1) {print OUTFILE ","};
        print OUTFILE  "\n";
        }
print OUTFILE "};";

close (OUTFILE);
```

■ **FIGURE 12.16**   PERL script for pseudo-random coefficient generation

```
module randnums;
  int ALPHA, BETA, GAMMA, DELTA;
  initial begin
    //pseudo-random 32 bit number
    ALPHA = $random;
    //pseudo-random 32 bit number with a seed
    BETA = $random(2);
    //pseudo-random number ranging from -9 to +9
    GAMMA = $random % 10;
    //pseudo-random number ranging from 0 to 9
    DELTA = {$random} % 10;
    $display ("ALPHA = %d, BETA = %d, GAMMA = %d, DELTA = %d", ALPHA, BETA, GAMMA, DELTA);
  end
  endmodule
```

■ **FIGURE 12.17**   Using $random to generate pseudo-random numbers

FIFO

Hierarchical code for the FIFO developed in Chapter 7 is shown in Figures 12.19–12.27. A small test fixture for the complete design is shown in Figure 12.28.

```
module svrands;
  int ALPHA, BETA;
  initial begin
    //set seed to 10 for deterministic sequence
    $srandom(10);
    //pseudo-random 32 bit unsigned number
    ALPHA = $urandom;
    //pseudo-random number from 2 to 25
    BETA = $urandom_range(2,25);
    $display ("ALPHA = %d, BETA = %d", ALPHA, BETA);
  end
endmodule
```

■ **FIGURE 12.18** SystemVerilog pseudo-random generator functions

```
/*Scalable, synthesizable FIFO design*/

module fifo_top #(WIDTH = 4, DEPTH = 8)
  (input WR_CLK, RD_CLK, RST, WR_EN, RD_EN,
  input [WIDTH - 1 : 0] DATA_IN,
  output [WIDTH - 1 : 0] DATA_OUT, output FULL, EMPTY);

  wire [$clog2(DEPTH) : 0] WR_BIN, RD_BIN;
  wire WR_AASD, RD_AASD;

  //Synthesizable memory array
  regfile MEM(WR_CLK, (WR_EN && !FULL), DATA_IN, WR_BIN[$clog2(DEPTH) - 1: 0],
    RD_BIN[$clog2(DEPTH) - 1: 0], DATA_OUT);

  //Write side reset synchronizer
  aasd WR_AA(WR_CLK, RST, WR_AASD);
  //Read side reset synchronizer
  aasd RD_AA(RD_CLK, RST, RD_AASD);
  //Write pointer counter
  counter WR_PTR(WR_CLK, WR_AASD, (WR_EN && !FULL), WR_BIN);
  //Read pointer counter
  counter RD_PTR(RD_CLK, RD_AASD, (RD_EN && !EMPTY), RD_BIN);
  //Hierarchical flag generator subsysem
  flags FLAGS(WR_CLK, RD_CLK, WR_AASD, RD_AASD,
      WR_BIN, RD_BIN, FULL, EMPTY);
endmodule
```

■ **FIGURE 12.19** Top level of hierarchical FIFO design

```
/*Data storage register file
One read port, one write port*/

module regfile #(WIDTH = 4, DEPTH = 8)
  (input WR_CLK, WR_EN, input [WIDTH - 1 : 0] DATA_IN,
  input [$clog2(DEPTH) - 1 : 0] WR_ADDR, RD_ADDR,
  output logic [WIDTH - 1 : 0] DATA_OUT);

  reg [WIDTH - 1 : 0] MEM [0 : DEPTH - 1];

  always_ff @(posedge WR_CLK)
    if (WR_EN) MEM[WR_ADDR] <= DATA_IN;
    else MEM[WR_ADDR] <= MEM[WR_ADDR];
  always_comb DATA_OUT = MEM[RD_ADDR];
endmodule
```

■ **FIGURE 12.20** FIFO register file

```
//Two stage reset synchronizer
module aasd(input CLK, RST, output logic SYNC_RST);
  logic STAGE1;
  always_ff @(posedge CLK, negedge RST)
    if (!RST) begin
      SYNC_RST <= 1'b0;
      STAGE1 <= 1'b0;
    end
    else begin
      SYNC_RST <= STAGE1;
      STAGE1 <= 1'b1;
    end
 endmodule
```

■ **FIGURE 12.21** Reset synchronizer

```
/*Generic, scalable binary counter.
It will be the log2 ceiling of parameter DEPTH bits.*/

module counter #(DEPTH = 8) (input CLK, RST, EN, output logic [$clog2(DEPTH) : 0]CNT);
  always_ff @(posedge CLK, negedge RST)
  if (!RST) CNT <= '0;
  else
    if (EN) CNT <= CNT + 1;
    else CNT <= CNT;
endmodule
```

■ **FIGURE 12.22** Generic counter used for address pointers

```
/*Hierarchical model of empty and full flag logic.
Full flag is write side, so uses write clock.
Empty flag is read side, so uses read clock.
Both sides use a two stage synchronizer on their
Gray bus inputs.
A Binary to Gray converter function is included to
convert the Gray code that crosses the clock domain
boundaries to binary for easy comparison of unequal
values.*/

module flags #(DEPTH = 8) (input WR_CLK, RD_CLK, WR_AASD, RD_AASD,
       input [$clog2(DEPTH) : 0] WR_BIN, RD_BIN,
       output FULL, EMPTY);
       wire [$clog2(DEPTH) : 0] RD_GRAY_SYNC, WR_GRAY_SYNC;

       sync WR_SYNC(WR_CLK, B2G(RD_BIN), RD_GRAY_SYNC);
       sync RD_SYNC(RD_CLK, B2G(WR_BIN), WR_GRAY_SYNC);

       full_logic FL(WR_CLK, WR_AASD, WR_BIN, RD_GRAY_SYNC, FULL);
       empty_logic EM(RD_CLK, RD_AASD, B2G(RD_BIN), WR_GRAY_SYNC, EMPTY);

       function [$clog2(DEPTH) : 0] B2G(input [$clog2(DEPTH) : 0] BIN);
       B2G[$clog2(DEPTH) ] = BIN[$clog2(DEPTH) ];
       for (int I = $clog2(DEPTH) - 1; I >= 0; I--)
         B2G[I] = BIN[I] ^ BIN[I + 1];
       endfunction
endmodule
```

■ **FIGURE 12.23** Hierarchical subdesign for flag logic

```
/*Scalable full flag design. Uses Gray to binary converts.*/

module full_logic #(DEPTH = 8) (input CLK, RST,
  input [$clog2(DEPTH) : 0] WR_BIN, RD_GRAY,
  output reg FULL);
  logic [$clog2(DEPTH) : 0] RD_BIN;

  function [$clog2(DEPTH) : 0] G2B(input [$clog2(DEPTH) : 0] GRAY);
        G2B[$clog2(DEPTH) ] = GRAY[$clog2(DEPTH) ];
        G2B[$clog2(DEPTH) - 1] =  GRAY[$clog2(DEPTH) ] ^  GRAY[$clog2(DEPTH) - 1];
        for (int I = $clog2(DEPTH) - 2; I >= 0; I--)
          G2B[I] = G2B[I+1] ^ GRAY[I];
  endfunction

  always_comb RD_BIN = G2B(RD_GRAY);

  always_comb
      if (WR_BIN[$clog2(DEPTH)] != RD_BIN[$clog2(DEPTH)] &&
        WR_BIN[$clog2(DEPTH) - 1 : 0] == RD_BIN[$clog2(DEPTH) - 1 : 0])
        FULL <= 1'b1;
      else
        FULL <= 1'b0;
endmodule
```

■ **FIGURE 12.24** Full flag generator

```
/*Scalable comparator. Returns true (logic one)
when two inputs are equal.*/

module empty_logic #(DEPTH = 8) (input CLK, RST,
  input [$clog2(DEPTH): 0] RD_GRAY, WR_GRAY,
  output reg EMPTY);

  always_comb EMPTY <= RD_GRAY == WR_GRAY;

endmodule
```

■ **FIGURE 12.25** Empty flag generator

```
/*Scalable two stage synchronizer.*/

module sync #(DEPTH = 8)(input CLK, input [$clog2(DEPTH) : 0] DATA,
  output logic [$clog2(DEPTH) : 0] SYNC_DATA);

  logic [$clog2(DEPTH) : 0] STAGE1;
  always_ff @(posedge CLK) begin
    STAGE1 <= DATA;
    SYNC_DATA <= STAGE1;
  end
endmodule
```

■ **FIGURE 12.26** Scalable synchronizer

```
module tb_fifo;
  timeunit 1ns;
  parameter WIDTH = 4;
  reg RD_CLK, WR_CLK, RST;
  reg WR_EN, RD_EN;
  reg [WIDTH - 1 : 0] DATA_IN;
  wire FULL, EMPTY;
  wire [WIDTH - 1 : 0] DATA_OUT;

  fifo_top UUT(WR_CLK, RD_CLK, RST, WR_EN, RD_EN, DATA_IN, DATA_OUT, FULL, EMPTY);

  initial $monitor("At %d, RD_BIN = %b, RD_GRAY = %b, WR_BIN = %b, WR_GRAY = %b",
   $time, UUT.RD_BIN, UUT.FLAGS.RD_GRAY_SYNC, UUT.WR_BIN, UUT.FLAGS.WR_GRAY_SYNC);

  initial begin
    RD_CLK = 1'b1;
    forever #4 RD_CLK = ~RD_CLK;
  end

  initial begin
    WR_CLK = 1'b1;
    forever #1 WR_CLK = ~WR_CLK;
  end

  initial begin
    RST = 1'b1;
    # 3 RST = 1'b0;
    # 12 RST = 1'b1;
  end

  initial begin
    RD_EN = 1'b0; WR_EN = 1'b0;
    #20 WR_EN = 1'b1; RD_EN = 1'b1;
  end

  always_ff @(posedge WR_CLK, negedge RST)
  if (!RST) DATA_IN <= 0;
  else if (WR_EN && !FULL) DATA_IN <= DATA_IN + 1;
  else DATA_IN <= DATA_IN;
endmodule
```

■ **FIGURE 12.27** FIFO test fixture

DMX RECEIVER

Figures 12.28–12.37 show a hierarchical implementation of a DMX512 receiver and test fixture as described in Chapter 7.

```
/*Top level of the DMX512 receiver.*/

module dmx(input CLK, RST, DATA_IN, input [8:0] SLOT,
  output wire [7:0] PACKET, output wire VALID);
  wire SYNC_PULSE, BIT_EN, FRAME_EN;
  wire [3:0] BITCNT;
  wire [8:0] FRAME;
  sync     sync(CLK, DATA_IN, DATA2);
  detect   d0(CLK, DATA2, SYNC_DATA, SYNC_PULSE);
  start_det sd(CLK, RST, BIT_EN, SYNC_DATA, START);
  hex_cnt c0(CLK, RST, SYNC_PULSE, BIT_EN);
  bit_cnt c1(CLK, RST, BIT_EN, START, SYNC_DATA, BITCNT, FRAME_EN);
  frame_cnt c2(CLK, RST, START, (FRAME_EN && BIT_EN && !SYNC_DATA), FRAME);
  data_reg r0(CLK, RST, SYNC_DATA, BIT_EN, FRAME_EN, BITCNT, FRAME, SLOT, PACKET, VALID);
endmodule
```

■ **FIGURE 12.28** Top-level DMX512 receiver

```
/*Two stage synchronizer.
The output will be synchronized to the receiving clock. This
type of synchronizer will ensure to a high probability
that indeterminate, metastable values will not propagate
but the output is not guaranteed to be any specific value.*/

module sync(input CLK, DIN, output reg DOUT);
  reg STAGE1;
  always_ff @(posedge CLK) begin
    STAGE1 <= DIN;
    DOUT <= STAGE1;
  end
endmodule
```

■ **FIGURE 12.29** Two-stage synchronizer for serial data

```
/*Start sequence detector.
A valid start sequence is at least 22 bits of logic zero
followed by at least 2 bits of logic one.
Once that sequence has been met, the next logic zero
will be the start bit of the first slot.
*/

/*Unusual case: suppose a start sequence is immediately followed
by another start sequence. In that case, the first logic zero
should cause ZCNT to go to one. Need to cover this potential event.
*/

module start_det(input CLK, RST, EN, DATA_IN, output logic START);
  reg [4:0] ZCNT; //Zeros count
  reg [1:0] OCNT; //Ones count
  always_ff @(posedge CLK, negedge RST)
    if (!RST) ZCNT <= 0;
    else //increment if DATA_IN == 0 and not yet at 22
      if (EN && START && !DATA_IN) ZCNT <= 0;
      else if (EN && DATA_IN && ZCNT < 22) ZCNT <= 0;
      else if (EN && ZCNT == 22) ZCNT <= 22;
      else if (EN) ZCNT <= ZCNT + 1;
      else ZCNT <= ZCNT;

  always_ff @(posedge CLK, negedge RST)
    if (!RST) OCNT <= 0;
    else
      if (EN && !DATA_IN) OCNT <= 0;
      else if (EN && OCNT == 2 && DATA_IN) OCNT <= 2;
      else if (EN && ZCNT == 22 && DATA_IN) OCNT <= OCNT + 1;
      else OCNT <= OCNT;

  always_comb START = (ZCNT == 22 && OCNT == 2);
endmodule
```

■ **FIGURE 12.30** Start sequence detector

```
/*A hexadecimal counter is the heart of the
locking algorithm. It will synchronously return
to zero whenever a transition is detected on
the input data stream. It outputs its terminal
count, which can be used to enable the bit
counter. Terminal count is set high when the
counter rolls over after reaching 15 or when
a data transition is detected and the count has
passed its midpoint.*/

module hex_cnt(input CLK, RST, SYNC,
  output logic TC);
  logic [3:0] COUNT;
  always_ff @(posedge CLK, negedge RST)
    if (~RST) COUNT <= 0;
    else
      if (SYNC) COUNT <= 0;
      else COUNT <= COUNT + 1;

    always_comb TC = (COUNT == 15 || (SYNC && COUNT > 7));

endmodule
```

■ **FIGURE 12.31** Locking counter

```
/*The frame counter determines where the input data stream
is in the frame in terms of time slots. Each receiving device
is enabled for just one time slot. The count output is compared
to the slot input to determine if the data being sent over the
common bus are for this device or some others.*/

module frame_cnt(input CLK, RST, CLR, EN, output reg [8:0] FRAME);
  always_ff @(posedge CLK, negedge RST)
    if (!RST) FRAME <= 0;
    else
      if (FRAME == 512 && EN) FRAME <= 0;
      else if (CLR) FRAME <= 0;
      else if (EN) FRAME <= FRAME + 1;
      else FRAME <= FRAME;
endmodule
```

■ **FIGURE 12.32** Frame counter

```
/*The detector finds transitions on the synchronized data
stream and sends out a synchronization pulse when the start
of a new bit is found. It operates at 16 times the nominal
data rate.*/

module detect(input CLK, DIN, output reg SYNC_DATA, SYNC_PULSE);
  always_ff @(posedge CLK) SYNC_DATA <= DIN;
  always_comb SYNC_PULSE = SYNC_DATA != DIN;
endmodule
```

■ **FIGURE 12.33** Transition detector

```
/*The data register is a 10 bit shift register that accepts the
serial data packet input and outputs a parallel byte and a validation
signal when the packet is determined to be correctly formatted.*/

module data_reg(input CLK, RST, DATA_IN, EN, FRAME_EN, input [3:0] BITCNT,
  input [8:0] FRAME, SLOT,
  output logic [7:0] PACKET, output logic VALID);
  reg [9:0] SHIFTREG;
  always_ff @(posedge CLK)
    if (BITCNT == 10)
      SHIFTREG <= SHIFTREG;
    else if (FRAME == SLOT && EN) begin
      SHIFTREG[0] <= DATA_IN;
      SHIFTREG[9:1] <= SHIFTREG[8:0];
    end
    else SHIFTREG <= SHIFTREG;

    always_comb PACKET = SHIFTREG[9:2];
    always_ff @(posedge CLK, negedge RST)
      if (!RST) VALID = 1'b0;
      else
        if (FRAME == SLOT && EN && BITCNT != 10)
          VALID <= 1'b0;
        else if (SHIFTREG[1:0] == 2'b11 && (FRAME == SLOT) && EN)
          VALID <= 1'b1;
        else VALID <= VALID;
endmodule
```

■ **FIGURE 12.34** Shift register and data validator

```
/*This counter tracks the bit position within a slot.
Because there may be an indeterminate number of logic one
bits received following the required two stop bits, the
counter holds its terminal count of 10 until the start of
a new slot causes it to return to zero.*/

module bit_cnt(input CLK, RST, EN, CLR, SYNC_DATA,
  output logic [3:0] CNT, output logic TC);
  always_ff @(posedge CLK, negedge RST)
    if (!RST) CNT <= 0;
    else
      if (EN && CNT == 10 && !SYNC_DATA) CNT <= 0;
      else if (EN && CLR) CNT <= 0;
      else if (EN && CNT < 10) CNT <= CNT + 1;
      else CNT <= CNT;
  always_comb TC = CNT == 10;
endmodule
```

■ **FIGURE 12.35** Bit counter

```
module tb_dmx;
  timeunit 1us;
  timeprecision 1ns;
  //Both periods should work for 3.92 <= period <= 4.08
  //They do not need to be the same. The capture range
  //must be +/- 2% for each.
  parameter PERIOD = 4.0;
  //Real is needed to prevent integer rounding when 16x clock is created
  parameter CLKPERIOD = 4.0;
  //STOPBITS can be any integer >= 2
  parameter STOPBITS = 3;
  //SPACE can be any integer >= 23
  parameter SPACE = 23;
  //MARK can be any integer >= 2
  parameter MARK = 4;
  reg CLK, RST, DATA_IN;
  reg [8:0] SLOT = 1; //UUT will be in slot 1.
  wire [7:0] PACKET;
  wire VALID;

  //The Unit Under Test
  dmx UUT(CLK, RST, DATA_IN, SLOT, PACKET, VALID);

  task sendframe;
    DATA_IN = 1'b0; //Start Space for Break
    #(SPACE * PERIOD) DATA_IN = 1'b1; //Start Mark for Break
    #(MARK * PERIOD);
  endtask

  task sendpacket(input [7:0] DATA);
    DATA_IN = 1'b0; //Logic 0 Start Bit
    #(PERIOD); //Send packet MSB first
    for (int I = 7; I >= 0; I--) begin
      DATA_IN = DATA[I];
      #(PERIOD);
    end
    DATA_IN = 1'b1; //stop bits
    #(STOPBITS * PERIOD) ;
  endtask
```

■ **FIGURE 12.36** Test fixture part 1

```
initial begin //16x Clock
  CLK = 1'b1;
  forever #(CLKPERIOD/32.0) CLK = ~CLK;
end

initial begin
  RST = 1'b1; DATA_IN = 1'b1; //Toggle reset
  #(3 * PERIOD) RST = 1'b0;
  #(3 * PERIOD) RST = 1'b1;
  sendframe;
  sendpacket(0); //Slot 0 data
  sendpacket(8'hAA); //Slot 1 data
  sendframe;
  sendpacket(8'hFF);
  sendpacket(8'h55);
  sendframe;
  sendpacket(8'h0);
  sendpacket(8'hDE);
end

initial begin
  @(posedge VALID)
  if (PACKET != 8'hAA) $display ("Error: Received %h instead of AA", PACKET);
  @(posedge VALID)
  if (PACKET != 8'h55) $display ("Error: Received %h instead of 55", PACKET);
   @(posedge VALID)
  if (PACKET != 8'hDE) $display ("Error: Received %h instead of DE", PACKET);
  end
endmodule
```

■ **FIGURE 12.37**   Test fixture part 2

SystemVerilog keywords

The following are the reserved words per IEEE Standard 1800. Although not all will be implemented in all design automation tools, none should be used for identifiers.

Verilog is case sensitive. To be recognized as a keyword, these words must be all lower case. The code in Figure A.1 uses capitalized keywords as identifiers. Any temptation to use this technique should be resisted.

```
/*The following code uses capitalized keywords as identifiers.
It does work. This technique is strongly discouraged.*/

module Module(input Input, Wire, output reg Output, Reg);
  always_ff @(posedge Wire) begin
    Output <= Input;
    Reg <= ~Input;
  end
endmodule
```

■ FIGURE A.1 Legal but awkward use of capitalized keywords as identifiers

accept_on	assert	bins
alias	assign	binsof
always	assume	bit
always_comb	automatic	break
always_ff	before	buf
always_latch	begin	bufif0
and	bind	bufif1

Digital Integrated Circuit Design Using Verilog and SystemVerilog 978-0-12-408059-1

byte	endgenerate	ifnone
case	endgroup	ignore_bins
casex	endinterface	illegal_bins
casez	endmodule	implements
cell	endpackage	implies
chandle	endprimitive	import
checker	endprogram	incdir
class	endproperty	include
clocking	endspecify	initial
cmos	endsequence	inout
config	endtable	input
const	endtask	inside
constraint	enum	instance
context	event	int
continue	eventually	integer
cover	expect	interconnect
covergroup	export	interface
coverpoint	extends	intersect
cross	extern	join
deassign	final	join_any
default	first_match	join_none
defparam	for	large
design	force	let
disable	foreach	liblist
dist	forever	library
do	fork	local
edge	forkjoin	localparam
else	function	logic
end	generate	longint
endcase	genvar	macromodule
endchecker	global	matches
endclass	highz0	medium
endclocking	highz1	modport
endconfig	if	module
endfunction	iff	nand

negedge	real	struct
nettype	realtime	super
new	ref	supply0
nexttime	reg	supply1
nmos	reject_on	sync_accept_on
nor	release	sync_reject_on
noshowcancelled	repeat	table
not	restrict	tagged
notif0	return	task
notif1	rnmos	this
null	rpmos	throughout
or	rtran	time
output	rtranif0	timeprecision
package	rtranif1	timeunit
packed	s_always	tran
parameter	s_eventually	tranif0
pmos	s_nexttime	tranif1
posedge	s_until	tri
primitive	s_until_with	tri0
priority	scalared	tri1
program	sequence	triand
property	shortint	trior
protected	shortreal	trireg
pull0	showcancelled	type
pull1	signed	typedef
pulldown	small	union
pullup	soft	unique
pulsestyle_ondetect	solve	unique0
pulsestyle_onevent	specify	unsigned
pure	specparam	until
rand	static	until_with
randc	string	untyped
randcase	strong	use
randsequence	strong0	uwire
rcmos	strong1	var

vectored	weak	with
virtual	weak0	within
void	weak1	wor
wait	while	xnor
wait_order	wildcard	xor
wand	wire	

Standard combinational and sequential functions

Digital circuits operate in binary. Signals have only two legal states, true or false, also called logic one and logic zero. The electrical characteristics of logic one and logic zero vary. At the dawn of digital electronic circuitry, when vacuum tubes were used to implement logic functions, the difference between a true and a false signal would be tens of volts. For modern semiconductors, it is less than 1 V. Use of the abstract true and false allows logic systems to be designed and built without specifying the electrical characteristics of each signal.

COMBINATIONAL FUNCTIONS

The most fundamental building blocks of all digital circuit are the Boolean operators AND, OR, Exclusive OR, and NOT. NOT can be combined with the first three to form NAND, NOR, and Exclusive NOR functions. The set of basic components also includes buffers, which do not change the logic flow but are often used to improve electrical and timing characteristics of a circuit.

These functions may be implemented as logic gates. The standard schematic symbols for each gate type and a truth table for each are shown below.

While any digital circuit can be made from nothing but two input NAND or NOR gates, using all the available functions is more efficient.

Except for the single-input NOT and buffer functions, all the gates can be expanded to an arbitrary number of inputs. The examples below only show two input gates.

AND

The output is true (logic one) if and only if all inputs are true (Figure B.1 and Table B.1).

■ **FIGURE B.1** AND gate symbol

Table B.1 AND truth able

Input 1	Input 2	Output
0	0	0
0	1	0
1	0	0
1	1	1

OR

The output is true (logic one) if any input is true (Figure B.2 and Table B.2).

■ **FIGURE B.2** OR gate symbol

Table B.2 OR truth table

Input 1	Input 2	Output
0	0	0
0	1	1
1	0	1
1	1	1

Exclusive OR (XOR)

The output is true (logic one) if an odd number of inputs are true, counting zero as an even number (Figure B.3 and Table B.3).

■ **FIGURE B.3** XOR gate symbol

Table B.3 XOR truth table

Input 1	Input 2	Output
0	0	0
0	1	1
1	0	1
1	1	0

BUFFER

The output follows the input (Figure B.4 and Table B.4).

■ **FIGURE B.4** Buffer symbol

Table B.4 BUF truth table

Input	Output
0	0
1	1

Inverter (NOT)

The output is the logical inverse of the input (Figure B.5 and Table B.5).

■ **FIGURE B.5** Inverter symbol

Table B.5 NOT truth table

Input	Output
0	1
1	0

NAND

The output is false (logic zero) if and only if all inputs are true (logic one) (Figure B.6 and Table B.6).

■ **FIGURE B.6** NAND gate symbol

Table B.6 NAND truth table

Input 1	Input 2	Output
0	0	1
0	1	1
1	0	1
1	1	0

NOR

The output is true (logic one) if and only if all inputs are false (logic zero) (Figure B.7 and Table B.7).

■ **FIGURE B.7** NOR gate symbol

Table B.7 NOR truth table

Input 1	Input 2	Output
0	0	1
0	1	0
1	0	0
1	1	0

Exclusive NOR (XNOR)

The output is true (logic one) if an even number of inputs are true, counting zero as an even number (Figure B.8 and Table B.8).

■ **FIGURE B.8** XNOR gate symbol

Table B.8 XNOR truth table

Input 1	Input 2	Output
0	0	1
0	1	0
1	0	0
1	1	1

THREE STATE CELLS

Some digital cells, in addition to being able to output logic zero and logic one values, can be turned off, or put into high-impedance mode, often written as Hi–Z. These are known as three state, or Tri-state®, cells. A symbol and truth table for such a Tri-state cell are shown below. Tri-state cells allow multiple outputs to be connected together. As long as only one driver is enabled at any given time, the drivers will not interfere with each other and the common output will be driven to either logic zero or logic one.

Tri-state cells may be inverting or noninverting. The enable signal may be active high or active low. In the first example shown here, the enable is active high and the cell is inverting. In the second, the enable is active high and the cell is noninverting. Active high enable means that when the enable signal is logic one, the input is passed through to the output. When the enable signal is logic zero, the device is in high-impedance mode. Conversely, active low enable means that the device is turned off when the enable input is logic one (Figures B.9 and B.10 and Tables B.9 and B.10).

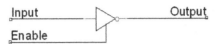

■ **FIGURE B.9** Active high enable inverting Tri-state buffer symbol

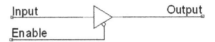

■ **FIGURE B.10** Active low enable noninverting Tri-state buffer symbol

Table B.9 Active high enable inverting Tri-state buffer truth table

Enable	Input	Output
0	0	Hi-Z
0	1	Hi-Z
1	0	1
1	1	0

Table B.10 Active low enable noninverting Tri-state buffer truth table

Enable	Input	Output
0	0	0
0	1	1
1	0	Hi-Z
1	1	Hi-Z

Half adder

An exclusive OR gate and an AND gate can be combined to form a half adder. Two half adder cells plus an extra OR gate can be combined to form a full adder, which is shown in Figure B.11 and Table B.11.

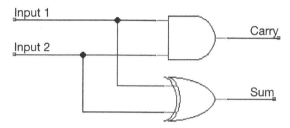

■ **FIGURE B.11** Half adder schematic diagram

Table B.11 Half adder truth table

Input 1	Input 2	Carry	Sum
0	0	0	0
0	1	0	1
1	0	0	1
1	1	1	0

Full adder

Full adders are a common building block of more complex arithmetic processing units. The gate level implementation shown below is one of many logically equivalent designs. It can be made from two half adders plus an additional OR gate (Figure B.12 and Table B.12).

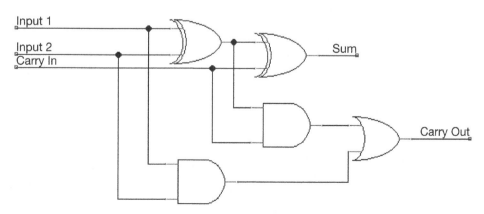

■ **FIGURE B.12** Full adder schematic diagram

Table B.12 Full adder truth table

Input 1	Input 2	Carry In	Sum	Carry Out
0	0	0	0	0
0	0	1	1	0
0	1	0	1	0
0	1	1	0	1
1	0	0	1	0
1	0	1	0	1
1	1	0	0	1
1	1	1	1	1

Multiplexor

Multiplexors select one of the two or more outputs. In the two bit example below, the output will be equal to A if Select is zero. If Select is one, the output will be equal to B. For larger multiplexors, more select bits are needed. A four-to-one multiplexor requires two select bits and an eight-to-one design requires three (Figure B.13 and Table B.13).

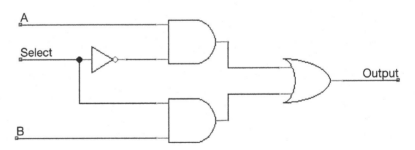

A
Select
Output
B

■ **FIGURE B.13** Two-to-one multiplexor schematic diagram

Comparator

Comparators check for equality between two buses. In the two-bit example below, Equal will be set true when A[0] = B[0] and A[1] = B[1]. Comparator can be made for buses of any width. The

Table B.13 Multiplexor truth table

A	B	Select	Output
0	0	0	0
0	0	1	0
0	1	0	0
0	1	1	1
1	0	0	1
1	0	1	0
1	1	0	1
1	1	1	1

gate-level diagram shown is not the only possible implementation of a comparator (Figure B.14).

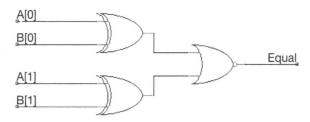

■ **FIGURE B.14** Two-bit comparator schematic diagram

SEQUENTIAL FUNCTIONS

The essential difference between a combinational function and a sequential one is that sequential functions can hold state. Sequential functions are further divided into level-sensitive latches and edge-triggered flipflops. Flipflops can also have level-sensitive control inputs, as is the case with ones that have asynchronous preset and/or reset inputs, but their normal operating mode is edge triggered.

RS latch

The RS, or Reset/Set, latch is not commonly found as a separate component. These latches are used as building blocks to form other sequential components, as shown in Figures 7.2 and 7.4. Examples of RS latch structures are shown in Figure 7.1 and their operation is shown in Table 7.1.

D latch

The D, or Data, latch is the simplest commonly used sequential cell. Its output follows its data input as long as a control signal, also called the gate, is active. When the gate is inactive, the output is latched, and any changes on the data input have no effect on the output.

Active may mean logic zero or logic one. D latches with both polarities are common. Table B.14 shows a D latch with active low control.

Table B.14 D latch operation table

D	Gate	Output
0	0	0
1	0	1
Any	1	No change: hold previous value

D flipflop (DFF)

The rest of the sequential functions are all edge triggered. Edges are indicated by arrows, $\uparrow$ for rising edges and $\downarrow$ for falling. All the examples given are sensitive to rising edges, which is the way the overwhelming majority of real devices work. Some, however, are activated on falling edges and ignore rising edges.

The DFF is an edge-triggered device similar to a D latch except the output only changes in response to an edge on the control signal, rather than allowing the output to immediately reflect any changes on the input as long as the control is active. Table B.15 shows the behavior of a rising-edge triggered flipflop.

Flipflops may also have asynchronous reset or preset inputs. When activated, these inputs set the output to logic zero or logic

Table B.15 D flipflop operation table

Clock	D	Output
↑	0	0
↑	1	1
↓	Any	No change: hold previous state
Steady state	Any	No change: hold previous state

one, respectively, regardless of the state of the clock or data inputs. When both are activated simultaneously, results are unpredictable. This is true of JK and T flipflops as well as D type. Preset and reset inputs are most often active low.

JK flipflop (JKFF)

JK flipflops have two data inputs. When both are asserted, the output toggles. When J alone is asserted, the output is synchronously set. When K alone is asserted, the output is synchronously reset. When neither is asserted, it holds state (Table B.16).

Table B.16 JK flipflop operation table

Clock	J	K	Output
↑	0	0	No change: hold previous state
↑	0	1	0
↑	1	0	1
↑	1	1	Toggle from previous state
↓	Any	Any	No change: hold previous state
Steady State	Any	Any	No change: hold previous state

Toggle flipflop (TFF)

A TFF changes state (toggles) with every active clock edge as long as its T input is true. T flipflops can easily be made from D or JK flipflops. To turn a JKFF into a T, simply connect the input to both the J and K ports of the flipflop. A T flipflop function made from a D flipflop and an XOR gate is shown schematically below (Figure B.15 and Table B.17).

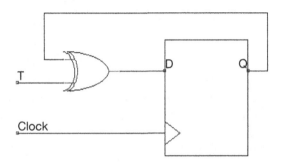

FIGURE B.15 T flipflop function implemented with D flipflop

Table B.17 T flipflop operation table

Clock	T	Current State	Next State
↑	1	0	1
↑	1	1	0
↑	0	Any	No change: hold previous state
↓	Any	Any	No change: hold previous state
Steady state	Any	Any	No change: hold previous state

Number systems

Digital circuits all work in binary, where each digit can only take the values of zero or one. Despite this seeming limitation, any arithmetic problem can be solved to any arbitrary level of precision, if enough resources are brought to bear.

In binary numbers, each digit represents a power of two. For a four-digit number, the maximum value will be 15, as the value will be $d_0 \times 2^0 + d_1 \times 2^1 + d_2 \times 2^2 + d_3 \times 2^3$ for digits d_0 through d_3. If all four are set (logic 1), then the value will be $1 + 2 + 4 + 8$, or 15.

Because binary numbers can only take two values per digit, it takes more digits to represent numbers than are necessary in the decimal system. The largest number that can be represented in n binary bits is $2^n - 1$. If negative numbers are also to be used, the largest value that can be represented in a given number of binary bits is approximately halved. Table C.1 shows decimal numbers for zero to 15, the same numbers in hexadecimal (Base 16), and their four-bit binary equivalent representations. Use of hexadecimal numbers provides a convenient way of grouping together four binary bits to form a more human-readable representation of binary numbers. Hexadecimal digits use the letters A through F to represent the values 10–15. Upper and lower case letters are both used.

The fundamental building block of computer arithmetic circuits is the full adder, which was shown in Appendix B. A full adder has three single-bit inputs, which it adds together to form a sum term and a carry out term.

Table C.1 Decimal, hexadecimal, and binary numbers

Decimal	Hexadecimal	Four-Bit Unsigned Binary
0	0	0000
1	1	0001
2	2	0010
3	3	0011
4	4	0100
5	5	0101
6	6	0110
7	7	0111
8	8	1000
9	9	1001
10	A	1010
11	B	1011
12	C	1100
13	D	1101
14	E	1110
15	F	1111

An adder, as the name implies, will add two numbers. Users of computers often want to subtract, multiply, and divide too, but no additional circuitry is needed to accomplish these other arithmetic functions.

Subtracting one number from another is the same as adding a negative copy of one to the other. This arithmetic equivalence is the basis for hardware subtraction. Multiplication can be accomplished by repeated addition and division by repeated subtraction. There are other methods of performing multiplication and division in hardware that offer greater performance than repeated addition and subtraction, but the fundamental building block of arithmetic units remains the adder.

In order to use negative numbers, some standard for their representation must be established. The most common standards are signed magnitude, one's complement and two's complement. While signed magnitude and one's complement are conceptually simple, almost all arithmetic circuits use two's complement representations.

Decimal	Four-Bit Binary Signed Magnitude
7	0111
6	0110
5	0101
4	0100
3	0011
2	0010
1	0001
0	0000
0	1000
−1	1001
−2	1010
−3	1011
−4	1100
−5	1101
−6	1110
−7	1111

Table C.2 Signed magnitude numbers

SIGNED MAGNITUDE

With signed magnitude numbers, the first digit represents the sign and the rest the magnitude, or absolute value. Some early digital computers worked entirely with signed magnitude binary numbers. It is still the way the mantissa field of floating point numbers is typically represented in computers. Table C.2 shows four-bit signed magnitude numbers. Like one's complement numbers, covered below, signed magnitude representation has the disadvantage of having two patterns that both mean zero.

ONE'S COMPLEMENT

To form the one's complement of a binary number, simply invert all bits. Thus for four-bit numbers, seven would be 0111 and negative seven 1000. Plus and minus seven are the largest and smallest numbers that can be represented in four-bit one's complement format. Four-bit one's complement numbers are shown in Table C.3.

A disadvantage of one's complement arithmetic is that it has two patterns that equal zero: for four-bit numbers, 0000 and 1111 both

Table C.3 One's and two's complement numbers

Decimal	One's Complement	Two's Complement
7	0111	0111
6	0110	0110
5	0101	0101
4	0100	0100
3	0011	0011
2	0010	0010
1	0001	0001
0	0000/1111	0000
−1	1110	1111
−2	1101	1110
−3	1100	1101
−4	1011	1100
−5	1010	1011
−6	1001	1010
−7	1000	1001
−8	N/A	1000

mean zero. Computer hardware implementing one's complement arithmetic can also be more complicated than some alternatives, as end-around carry is needed to adjust the final answer when an operand is negative. An example of end-around carry is shown below. In that example, negative one is added to four. The answer should be three. Without end-around carry, the result will be two. The carry bit needs to be added back to the sum for the correct answer of three to be obtained.

1110 (negative one, one's complement)
+0100 (plus four)
1 ← 0010 (positive two with a carry out. The answer is wrong and needs to be adjusted with end-around carry.)

TWO'S COMPLEMENT

The overwhelming majority of modern computers and other digital hardware use two's complement numbers. Two's complement numbers may be formed by first forming a one's complement and then adding one.

This adjustment eliminates the problem of double representation of zero. With no duplicate bit patterns, the range of two's complement numbers also increases by one. The range of two's complement numbers is from $2^{(n-1)} - 1$ down to $-2^{(n-1)}$ for n bits. Table C.3 shows one's and two's complement representations for decimal numbers 7 down to -8, the range that can be represented with four binary bits. Four-bit one's complement format does not have any way to represent -8.

ONE HOT

All the previous number systems utilize all possible bit patterns. This is most efficient in terms of bit utilization, but bit utilization is not always the most important criterion for digital circuit designers. One-hot encoding requires one bit for every number, which is exponentially more bits than are required for binary encoding. Binary encoding requires $\log_2 n$ bits to count from 0 to $n - 1$.

The benefit of one-hot encoding is that the values are predecoded. This feature is most useful in state machines, where different subsystems are enabled for each value of the state vector. The savings are in speed. Rather than create select lines by evaluating the state, enable signals are connected directly to the relevant bits of the one-hot encoded machine.

A sample one-hot encoding for numbers zero through seven is shown in Table C.4.

Table C.4 One-hot encoding

Decimal	Binary	One Hot
0	000	00000001
1	001	00000010
2	010	00000100
3	011	00001000
4	100	00010000
5	101	00100000
6	110	01000000
7	111	10000000

ONE COLD

One-cold encoding has the same advantages and disadvantages as one-hot. The difference is that the active bit is logic zero rather than logic one. One-cold encoding is most useful for systems that have active-low enable signals.

Table C.5 shows one-cold encodings for decimal values zero through seven.

Table C.5 One-cold encoding

Decimal	Binary	One Cold
0	000	11111110
1	001	11111101
2	010	11111011
3	011	11110111
4	100	11101111
5	101	11011111
6	110	10111111
7	111	01111111

Table C.6 Gray code

Decimal	Binary	Gray
0	0000	0000
1	0001	0001
2	0010	0011
3	0011	0010
4	0100	0110
5	0101	0111
6	0110	0101
7	0111	0100
8	1000	1100
9	1001	1101
10	1010	1111
11	1011	1110
12	1100	1010
13	1101	1011
14	1110	1001
15	1111	1000

GRAY CODE

Gray code is an alternative numerical representation that offers the advantage of having only one bit change at a time when incrementing or decrementing. Gray codes are only possible for even numbers and are only commonly used for sequences that are a power of two in length (Table C.6).

Index

Printed in the United States
By Bookmasters